FROM FALLING APPLES TO THE UNIVERSE

FROM FALLING APPLES TO THE UNIVERSE

A GUIDE FOR NEW PERSPECTIVES ON GRAVITY AND GRAVITATION

BY JOHN R. LAUBENSTEIN

Charleston, SC
www.PalmettoPublishing.com

*From Falling Apples to the Universe:
A Guide for New Perspectives on Gravity and Gravitation*

© 2021 by John R. Laubenstein

All rights reserved.

This book or any portion thereof may not be reproduced or used in any manner whatsoever without the express written permission of the publisher except for the use of brief quotations in a book review.

Hardcover ISBN: 978-1-64990-597-0
Paperback ISBN: 978-1-64990-825-4
eBook ISBN: 978-1-64990-138-5

TABLE OF CONTENTS

Preface   ix

Section 1: Understanding Gravity
and Gravitation

1. Gravity, Our Starting Point   3
2. Idea One: Force   7
3. Idea Two: General Relativity   11
4. Idea Three: Scale Metrics   15
5. Space: What the Heck Is It?   19
6. A Changing Metric from Gravitation   23
7. Space: A Three-Dimensional Entity   29
8. The Energime   33
9. Quantifying the Energime   37
10. An Isolated Island Economy   39
11. The (Inverse) Fine-Structure Constant   43
12. Electrostatic Force   49
13. A Quick Review   55
14. A Quantitative Look at GR versus Scale Metrics   59
15. Kinetic Energy   65
16. The Energime: Is It the Smallest Photon?   75
17. Momentum   79
18. The Equivalence Principle   87
19. Scale Metrics Equivalence Principle   95
20. Further Development of General Covariance   105
21. Black Holes: Are They Real?   111

22. "I Don't Believe You: We Have Seen Black Holes"..... 115
23. Local versus Global Geometry 121
24. The Circle... 125
25. An Apple Falling from a Tree Branch................. 129
26. Celestial Mechanics 133

Section 2:
Understanding the Universe

27. A Changing Metric with Time 143
28. Space Revisited.. 149
29. A Brief History of the Standard Model................ 157
30. The Scale Metrics Pictorial of the Universe 167
31. A Story of Three Journeys 171
32. Understanding Your Square of the Universe.......... 177
33. An Expanding and Cooling Universe? 183
34. A Local Perception of Space 187
35. A Big Bang?... 191
36. Matter...It's No Simple Thing 193
37. The Electron 197
38. The Proton 203
39. The Neutron... 207
40. A Little More on 137................................. 211
41. An Ever-Changing Universe 217
42. Redshift.. 219
43. Age of the Universe.................................. 223
44. Inflation—Is It Needed?.............................. 229
45. Dark Matter and Dark Energy 231
46. Is the Rate of Expansion Accelerating?................ 235
47. Is Space Stretching or Not? 243
48. A Not-So-Hot Beginning 249
49. Conclusion... 255

Section 3:
Understanding the Mathematics

50. Introduction 261
51. Planck's Constant as a Distance...................... 263

52. The Relationship between G and Planck's Constant.... 267
53. Momentum: Deriving the Scaling Factor............... 271
54. The Structure of Matter over the Ages................. 275
55. Explaining the Expansion Rate of the Universe 281
56. The Relationship between Kinetic Energy and Velocity
    Squared... 285
57. Implications for Gravitational Redshift from an Exact
    Kinetic Energy Equation 291
58. Resolving the Difference between GR and Scale Metrics
    Redshift Equations................................... 297
59. Why the Event Horizon Never Forms.................. 303
60. Relativity: A Cheat on Science 309

Additional Reading.. 327
About the Author 329
Index 331

PREFACE

This book is for readers who ponder the universe and question how it may have come to be the way it is. It is for those with an open mind who are willing to join me on a journey to explore new possibilities. The book provides a new perspective on gravity and gravitation and is written in a style that is intended to hold the interest of both more general readers as well as those that are experts in the field.

The first two sections, "Understanding Gravity and Gravitation" and "Understanding the Universe," are primarily written in a narrative style with some basic equations. Section 3, "Understanding the Mathematics," provides a more rigorous mathematical treatment that supports and defends the narrative in sections 1 and 2.

I have done this because I have often shared my writings with non-math-oriented associates and then asked the question, "Did you get through the math OK?" Ultimately, the answer is almost always the same: "It was fine, I had no problem with the math. I just skipped over those sections."

So the goal is to break out the heavy mathematics so as to not disrupt the flow of the narrative. But a word of caution for those that love the mathematics: don't skip the words. Sections 1 and 2 stand on their own merit using language to fully develop many new ideas. And an invitation to the more general reader: don't be scared away. If you have an interest in relativity, gravity, and the universe, this book may very well be for you.

It is my hope that *From Falling Apples to the Universe* provides all readers an interesting, compelling, and challenging new perspective on gravity and its role in our universe.

SECTION 1
UNDERSTANDING GRAVITY AND GRAVITATION

1

GRAVITY, OUR STARTING POINT

Why does an apple fall from a tree branch to the surface of the earth? This is the starting point of our journey—a question that we will revisit often, and one that will lead to new and far-reaching ideas.

There have been various explanations for gravity over the centuries, but the most relevant ones for this discussion are the ideas of Sir Isaac Newton and Albert Einstein along with the concepts of geometry developed by Euclid and Bernhard Riemann.

Let's start with some basics:

1. All good science starts with observation.
   a. If we drop an object, it will fall.
   b. It doesn't seem to matter much where we drop an object from; anywhere around the earth provides the same observation.
   c. It seems that the higher we drop an object (or the longer it falls), the faster it is moving when it hits the ground.
   d. It seems like the farther an object falls, the more energy it gives up when it hits the ground.
   e. And a very interesting one: if we drop two objects of different masses from the same height (and ignore air resistance), they both fall and hit the ground at the same time.
2. We then often look for explanations or rules that can be applied to explain what we see and, better yet, predict what might happen in future situations.

FROM FALLING APPLES TO THE UNIVERSE

    a. If I repeat any of the experiments from above, I get the same result. So these become my rules from which I can predict future outcomes.

    b. If an egg falls off a counter and breaks on impact, I can predict that if I repeat that experiment, the outcome will likely be the same: the egg will break.

3. We then try to refine these observations in a quantitative manner by using some form of mathematics.

    a. This is where language begins to fail us because terms like "fast," "slow," "very fast," "heavy," "light," and so forth are not extremely helpful beyond providing a relative comparison.

    b. For example, if I conduct several experiments with falling balls, I might record the following data. The ball in the first experiment hit the ground moving faster than it did in the second experiment, where the ball was dropped from a lower height. This may all be true, but our ability to formulate specific rules for the behavior of falling objects is limited if we cannot quantify our statements with mathematics. I think everyone can see the advantage of saying that the ball hit the ground moving thirty meters per second when dropped from forty-six meters but about twenty meters per second when dropped from twenty meters.

    c. We can achieve the above values by using the equation $V^2 = 2ad$, where V is the velocity, a is the rate of acceleration, and d is the distance fallen.

    d. So this is where Newton made a big contribution to science, in that he started to quantify and use mathematics to explain motion. Although Newton is often credited for this, it is important to note that many others over the centuries have also contributed to our quantitative understanding of motion.

4. Lastly, we attempt to formalize our observations and mathematics into some form of principle, concept, law of nature, or theory that provides some physical meaning to what we have observed and studied.

a. This is the tough part because even though we have quanti-fied some of our observations and turned them into math-ematical equations, we still are no closer to answering the question: Why does an apple fall to the ground? We can explain what happens but not why.

b. Note that this is not a required step. Sometimes the best we can do is know what will happen without necessarily answering why. But it is always better to provide the reason or cause. Your ability to predict outcomes is always stron-ger if you can explain why something happens as opposed to simply knowing what will happen.

However, one must also guard against thinking that if we can answer all the whys, we will then know everything. Even if we could, it is not that simple. There are no theories that are completely right or wrong. Science is not that black and white, despite what people (and some scientists) would like you to think. We have only ideas, concepts, and models that are developed by humans based on observations that per-haps can provide useful information. That is as far as science can take us. While science may be the search for reality, it seeks a goal it will never achieve. We must be satisfied with models that mimic and simulate reality.

So with that understanding, what are the models that might be useful in thinking about why an apple falls to the surface of the earth?

2

IDEA ONE: FORCE

The concept of a force—this is what most of us were taught in school. It is the simplest way to make sense of why an object falls. Gravity is a force that pulls objects to the surface of the earth. If an object speeds up as it falls to the surface, the earth must be exerting some type of force. It seems to make intuitive sense, right? A simple equation for force is

$$F = ma.$$

Where F is the force, m is the mass of the object on which the force is being exerted, and a is the resulting acceleration of the object due to the force applied to the object. The acceleration can be defined as a change in the object's velocity over a period of time.

When we use an equation in this form, we are implying that the acceleration remains constant over the entire time of our observation. This is not always the case in real-life situations; for example, the rate of acceleration changes as you fall from a great distance to Earth. So scientists prefer to use a derivative form, which can be thought of as the instantaneous force, or the force at any specific moment. That equation is

$$F = \frac{dp}{dt}.$$

Here, dp represents a tiny change in linear momentum (called an infinitesimal), and dt is the tiny interval of time over which the change in momentum occurred. I bring this up only because some more advanced readers may suggest that this is a better approach. Using this form of the equation, a scientist can utilize calculus to better understand what is occurring in a real-life situation where the acceleration is not constant for the entire duration being considered. For the most part, we are not going to be using calculus within section 1 of this book. Therefore, most equations will be expressed in their classical form with the understanding that there are limitations to how these classical equations can be used.

With that out of the way, there is a much bigger topic to discuss—that is, whether we even all agree on the meaning of the term "acceleration." I am sure that if you ask a scientist about acceleration, he or she will tell you that it is well defined, but the problem is that it is well defined within the context of how the scientist views acceleration. Is it the result of a force making something move in a way that it does not want to move? That is certainly one way to define it, and it would seem to be what is implied by the equation $F = ma$. Or is it simply an observation that an object's velocity is changing? This is certainly the approach used when defining acceleration as a change in velocity over time. In Euclidean space (that is, the flat geometry that we experience in everyday life), in the absence of gravity or other force fields, these two statements have the same meaning. The only way to change an object's velocity over time is to exert a force upon it, so a force is always involved with a change in velocity, and a force is always felt.

Force fields are a bit more abstract. For example, the force field (the force at various points) exerted by gravity is written as

IDEA ONE: FORCE

$$F = \frac{GMm}{r^2}.$$

The above equation is also a contribution to science made by Newton, where F is the force exerted by gravity, G is the gravitational constant, M is the mass of the gravitating body, m is the mass of the object on which the force is exerted, and r is the distance of separation between m and M. (To be technically correct, both the gravitating body and object exert a force on each other, but when the masses of the objects are very different—for instance, the mass of Earth compared to the mass of an apple—we can safely conclude that it is the apple falling to the surface of the earth.)

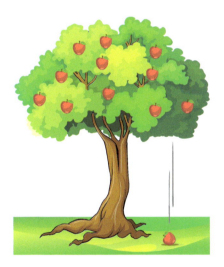

What is peculiar about this type of force is that you do not feel it if you are the one falling within it (if you ignore air resistance). What is exhilarating about a roller coaster is that as you fall downward through a big drop, you lose the sensation of a force acting on you. That funny feeling in your stomach is the absence of a force. A person falling toward the surface of the earth speeds up but does not feel it.

This is something that Albert Einstein recognized and used to help develop his idea of general relativity (GR).

So what do we mean by "acceleration"? Is it a change in velocity over time, or is it the sensation of a force causing something to move in a direction that is not its natural and preferred path of motion? These are human-made definitions, so there is no right or wrong answer, but we should at least acknowledge that a distinction needs to be made and pick one of them. For our purposes, acceleration is always felt. It is the result of forcing an object to move in a way that it does not naturally want to go. As for an object that has a changing velocity (such as the person in free fall), we will simply call that a changing velocity over time—but not acceleration. Please make a mental note of this because we will be returning to this often in future chapters.

There are other problems that emerge when defining gravity as a force. For starters, what is it about Earth (or any massive body) that exerts a force in the first place? And even if you understood how Earth exerts a force, there is a problem in that gravitational force appears to suggest an action at a distance. That is, two faraway objects instantaneously feel a force between them, as determined by Newton's equation for gravitational force.

It's also strange that the amount of force exerted is related to the mass of the object being attracted. Earth is Earth; it's one size and doesn't change, so you might think it exerts one single constant force on all objects at a given distance. But in actuality, the amount of force applied to two different objects at the same distance is different and depends on the mass of the falling object. Seems a bit forced (pun intended).

3

IDEA TWO: GENERAL RELATIVITY

E instein looked at this another way: maybe there is no force acting on an object as it falls. But if the object's speed is changing, how can that be? Einstein suggested that the geometry of space and space-time (the idea that space and time are interconnected within a four-dimensional coordinate system) may not be as simple as we think. Perhaps space-time is curved, and a falling object is not experiencing any force but moving exactly as it wishes along a geometry that is curved. This preserves Newton's first law and extends it (or, as scientists like to say, generalizes it) to include curved geometries.

This is a bit hard to visualize. After all, an apple falls straight down to the surface of the earth; how is that following a curve? The answer lies in the fact that you are only considering the space the apple is moving through and not the time component. Once you introduce time as a dimension, you'll see that the apple actually follows a curved path as it falls toward the earth.

But what's interesting about this curve is that it requires you to change your perception of geometry. It does not represent a curve in Euclidian space but rather suggests that the actual geometry of space-time is itself curved.

We think of Euclidean space as flat. Some of the properties of Euclidean space are that the three angles of a triangle

FROM FALLING APPLES TO THE UNIVERSE

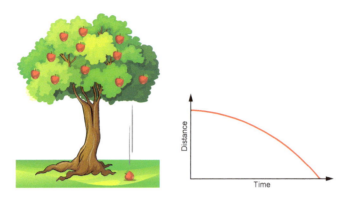

will always equal 180 degrees and that two parallel lines can be extended to infinity without crossing each other. As mentioned earlier, this is the geometry that we use in everyday life. There are actually five postulates that define Euclidean geometry, but you get the idea.

There are different notions of geometry. The one used by Einstein was Riemannian geometry, named after its developer, Bernhard Riemann. In the simplest sense, you can think of it this way: Imagine yourself in a room. You would typically think of the floor in that room to be flat. You can then extend those dimensions out to infinity, and you would have a flat surface that obeys the rules of Euclidean geometry. Now consider yourself in that same room but also realize you are on the surface of the earth, which is spherical. That does not change your perception that the floor is flat (in your local surroundings), but you realize that you cannot extend the floor's flat geometry to infinity and expect it to describe the geometry of the earth's surface. Riemannian geometry is a way of taking all the locally flat surfaces and combining them in a way that makes the larger (global) geometry curved. It provides the instructions on how to connect locally flat geometries to make a curved geometry. Note that this is not a curved surface in flat space; the model suggests that the geometry itself is actually curved. This is

IDEA TWO: GENERAL RELATIVITY

a subtle but important distinction that is somewhat hard to grasp. Our minds immediately want to trace a curve within the notion of a flat space. This is not how GR works. Within the presence of a gravitating mass, it is actually the geometry of space and space-time that is curving.

This leads to the concept of the metric (not the metric system of measurement but a different use of the word "metric.") The metric can be thought of as the shortest distance between two adjacent points in space. When we include time, the metric becomes the shortest distance between two adjacent events. An event simply includes the time associated with a position in space. An often-used example is that if I invite you to a party and provide you with my position (my address), it is not very helpful unless you know what time you should arrive. We think of time in terms of years, months, days, and hours. But scientists often use one continuously running stopwatch that started running at the beginning of the universe and has never been shut off.

In Euclidean geometry the metric is a straight line—that is, there is not a more direct path between two points than a straight line. But if my geometry is a sphere and I want to connect two points on its surface, I have to travel along an arc (a curve) to get from one point to the other. Regardless of how curved the geometry of the surface is (that is, how large the sphere is), the arc will always be longer than a straight line. Yet there indeed exists a shortest distance to connect the two points along the surface of the sphere. This shortest distance lies on what is called a geodesic, and it generalizes the idea of a straight line (shortest distance) in Euclidean geometry to the concept of the straightest possible path (shortest distance) on a curved geometry. Let's think about this: How can a curve be thought of as straight? Because it is the geometry of the space-time that is curving. You cannot think of this in terms of a curve in Euclidean (flat) space.

Motion is then related to how the value of the metric changes from place to place, and I can now explain why an object falls without having to impose a force. Rather, it is the curvature of the geometry of space-time that causes an apple to fall to the earth, and the value of the metric provides me with information on how fast the apple will fall. As suggested by the physicist John Wheeler, matter tells space-time how to curve and space-time tells matter how to move (or, in our case, fall). Note that we observe the apple to be falling, but within four-dimensional space-time, it is simply moving along a geodesic.

We can determine motion from how the metric changes, yet we need to ask what the metric is changing relative to. This introduces a somewhat confusing but perfectly legitimate concept of distance (as well as time and mass). If you are on a curved geometry, you are not aware of it locally. The measurement of distance in your local surroundings is called the proper distance. The way it would be measured if the space-time were not curved (that is, from a location far away, not near a gravitating mass) is called the coordinate distance. The ratio of the proper distance over the coordinate distance gives you information on how the metric has changed and why the apple will fall to surface of the earth.

This gives us a theory as to what is happening that was not available in Newton's idea of a force. Remember, there was no explanation as to what the force was that acted mysteriously and instantaneously between any two objects.

In Einstein's approach, a mass curves space-time (and not instantaneously; rather, the curvature is communicated outward at the speed of light). So this is an improvement over Newton. And what I mean by improvement is that it is a model that provides additional useful information that was not available using Newton's idea of a force.

But we still do not know why the space-time is curved by the mere presence of a mass.

4

IDEA THREE: SCALE METRICS

We need to be careful not to take curved space-time too literally. This is a mathematical model that provides us with information that agrees with observation. That is good, but it is not the same as saying we now understand the reality of the universe. We do not know if we live in a four-dimensional space-time that is curved by mass. We only know that this model provides some meaningful results. Sometimes all of us (even scientists) lose sight of this and begin to think far too literally about the perceived physical reality of the four-dimensional space-time continuum. It is often presented as a foregone conclusion, and it is an ongoing obsession of science fiction literature and movies. But that does not make it so.

The question I would ask is whether there is a model that can describe gravity using only the three dimensions that we are physically aware of. In addition, we tend to view these dimensions as Euclidean, so the goal is to explain gravity without using a force yet within the framework of three-dimensional Euclidean geometry. Some may call this a foolish attempt, but I believe this to be a worthwhile goal. Why shouldn't we be able to explain something within the boundaries of how we actually observe the world? It would certainly make gravity easier to understand, and that should be the goal of any model.

So what might this look like?

Just as in GR, if we are going to use something other than a force to explain gravity, then the notion of a metric would be very helpful—that is, the concept of the shortest distance between two adjacent points (or events) within the three dimensions that we are physically aware of. And by understanding how the metric changes locally as compared with the value of the metric far, far away (at infinity), we can determine why an apple falls to the surface of the earth.

So how might a metric change if the distance between two points is always measured using flat Euclidean geometry? The answer: the basic geometry does not change (that is, the geometry of space-time does not curve); rather, it's the physical scale of the dimensions that changes. For example, I can have various versions of the exact same map by changing the scale on the map. One map might be one inch to one mile, while another version of the exact same map might be one inch to five miles, and on and on we could go, displaying the same thing with maps using any number of different scales.

Now in this idea, when you move in the absence of a gravitating mass, you essentially stay on the same map using a constant scale. But when gravitation is introduced, your scale changes as you move radially inward toward the gravitating mass (a fancy way of saying you fall straight down toward the surface of the earth). So the third idea uses the concept of a changing scale in flat space rather than a changing geometry in curved space-time. It can easily be described as a theory of scale metrics. The metric is determined by the scale of the map, and the change in the metric is related to the scale of the map that you are locally present in as compared with the scale of the same map at infinity.

Now how does this determine the way an object falls? The value of your local metric and the value of the metric at infinity can be used in conjunction with the Pythagorean theorem $(a^2 + b^2 = c^2)$ to determine the velocity squared of an object as it falls from infinity to a specific location relative to the gravitating mass.

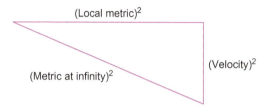

This may sound too simplistic, but in fact, even in GR the metric is a generalized version of the Pythagorean theorem in four dimensions using curved geometry. The mathematics of GR becomes quite complex, but the concept is similar to that of scale metrics. And, though rarely described this way, you can apply GR in the exact same manner as I have in scale metrics to determine the speed of an apple falling to the earth!

So we have two very similar approaches. In scale metrics we use a change in scale to change the value of the metric, and in GR we use the curvature in the geometry of spacetime to vary the metric. I believe that scale metrics provides a much simpler approach. But, perhaps more important, it can provide a physical explanation as to why the metric changes. This is exciting new information not available to us through either Newton's or Einstein's models of gravity.

Scale metrics can answer the question: What is it about the mere presence of a mass that causes the scale of space to change? We will be exploring this in upcoming chapters, but first we need to understand space.

5

SPACE: WHAT THE HECK IS IT?

I ask you to look at space in a totally new way. If you take nothing else away from this book, at least walk away with the notion that our current perceptions of space may be completely wrong. Space is fundamental to our understanding of gravity, gravitation, and the nature of the universe. It is the entity from which distance, mass, and time are derived, and it is critical to the understanding of matter, motion, and the formation of the universe. Put bluntly, space is far more complex and robust than we have ever given it credit for.

One of the major problems in cosmology is that we have no good definition for the concept of space. It is something we all experience and intuitively feel, but it lacks a truly rigorous definition. It is often described as the interval between two objects (but that is at best a definition of distance—not space). Also, space itself is said to have a vacuum mass that pops in and out of existence faster than it can be measured. So the idea of space becomes somewhat confusing. How can space be the interval between masses if space itself has mass-like properties? And how do we take a concept of distance and translate that to three dimensions? Lastly, why three dimensions, as opposed to two or four or any other number?

Scale metrics provides a definitive and quantitative explanation of space that may begin to answer some of the above questions. It is a rather simple concept. Ready?

Matter emits space.

In other words, matter undergoes a phase change from a bound state (which is present in matter) to an unbound or free state (representing space). Think of it like the phase change that water undergoes as it changes from a liquid to a gas. Or, better yet, the process of sublimation when dry ice goes directly from a solid to a gas.

The inside of the container has a small piece of dry ice that is producing a large volume of gas.

A piece of dry ice does not take up a lot of space. But the same number of molecules in the form of a gas takes up a significant amount of space. So think of the bound state as a solid (perhaps so compact that it approaches a singularity, meaning it would take up no space at all). This bound state undergoes, at a regular rate, a phase change to free space, which in turn defines the amount—or quantity—of space in the universe.

It is like blowing up the universe just as you would blow up a balloon. No analogy is perfect, but what I like about this one is that it does not matter what size balloon you start with. What determines the volume of any balloon is the amount of air blown into it. This is perhaps exactly the way it is with space!

This is not an abstraction but a literal suggestion. The scale metrics model suggests there is an actual particle present in both matter and space. It is the same particle, but it can exist in different states: bound (as it is in matter) or free (as it is in space). The distance between two free particles defines the metric. In this case the metric is physical, unlike in GR, where the metric is an abstraction. That is, the concept of the distance between two adjacent points of space has no real physical meaning in GR, yet in scale metrics it is real and physically present.

Let me remind you that this is merely an idea, one that will be developed into a model. So let's give it the opportunity and see if we can learn anything of interest from it. As stated at the conclusion of the previous chapter, scale metrics can be a useful model in understanding what it is about the mere presence of a massive body that causes the metric to change. Remember, for all the achievements of GR, it is unable to explain why a mass curves space-time. The GR model predicts what will happen, but it is unable to say why.

6

A CHANGING METRIC FROM GRAVITATION

The search for why the metric changes in the presence of a mass, from the scale metrics perspective, is related to the work of Ernst Mach. This is the same Mach whose name is used to refer to the speed of sound (for example, a rocket moving at Mach 1 or Mach 2). Mach had a very interesting idea on how the universe itself may relate to that which we observe locally. Mach's principal is somewhat difficult to express. In its most general sense, it states that the large-scale structure of the universe influences what happens locally. Einstein considered Mach's principle to be important in the conceptual development of GR; however, ultimately it was not included in the final mathematics of GR. This may be partly because it is very difficult to express Mach's principle in a mathematical form.

In GR, the mathematics that determines the curvature of space-time is influenced only by the mass of a gravitating body. The local mass tells space-time how to curve. In static time, the change in proper distance versus coordinate distance for a radial path inward (again, a fancy way of expressing the length of the metric as you move to different positions directly toward the center of the gravitating mass) is influenced only by the mass of the local gravitating body. It is possible to suggest that a different structure to the universe might provide a different value to the gravitational constant, but even acknowledging that, GR has no way of suggesting how the gravitational constant would change as the large-scale structure of the universe changes. Rather, in both Newton's gravitational force and Einstein's curved space-time, the focus of gravity is placed solely on the shoulders of the gravitating mass (and the distance that an object is from this mass).

Why might Mach's concept of large-scale structure be important to consider? One of the difficult things for all humans to comprehend is the magnitude of the universe. Why is it so darn big? Whether created by God or evolved through natural processes, it just seems to be overkill. Even if there were to be other life in the universe, it would appear to be so far from us that we could all live happily in a universe many times smaller. This also gives way to the idea that we are insignificant within the context of the entire universe, the feeling that we have little meaning within the overall picture.

Wouldn't we be perfectly happy if we just had our solar system (and maybe a few stars for the night sky)? Or, certainly, we would be content with the Milky Way galaxy. So why so much? Why such a large, overwhelming universe?

Scale metrics may have something to say about this. Both Newton and Einstein focus on the gravitating mass, but scale metrics takes into account the gravitating matter as it relates to all the mass (and energy) in the universe.

We assume (and Newton and Einstein perpetuate this idea) that gravity on Earth would be essentially the same if the universe were a different size, such as if it were limited to the Milky Way galaxy or even our solar system. The mass of Earth would be the same, and therefore it would provide the same instructions on how space-time should curve, according to Einstein. Or, using Newton's idea, Earth would provide the same force as determined solely by its mass (and the distance an object is from it).

But this may not be the case. If it were only the Milky Way galaxy that existed, the influence of gravity on Earth's surface might be much stronger. Scale metrics suggests that a 125-pound person on the surface of Earth would weigh roughly three billion pounds on the very same planet if the universe where limited to the Milky Way galaxy. If we were limited to only our solar system, that same person's weight would be more than four quadrillion pounds. Of course these are somewhat fictitious results because it is unlikely that the Milky Way galaxy or our solar system could form in its present state without the support of the large-scale structure of the universe. The point is that you cannot separate Earth, the solar system, or the Milky Way galaxy from the universe and expect everything to remain the same. The local properties of Earth are not independent of the scale of the universe. So perhaps there is a reason why the universe is so very large—so that we can experience things just the way we do here on Earth.

How does scale metrics come to this conclusion? It has to do with how the scale of each map changes as you get closer to the surface of Earth. If you think of all the matter of the universe as emitting space (through a phase change), then if the distribution of matter (and energy) throughout the universe is—at least on a large scale—uniform, then the metric of the entire universe is constant. It more or less looks like a grid, and this suggests a flat (or Euclidean) geometry.

Explaining gravity using Euclidean geometry, as you may recall, is one of my goals. But, locally, if you have a large mass, it will be emitting space from its center outward. This does not result in a curvature of space-time, but rather it is

Uniform metric of space from balanced distribution
of mass and energy

simply a change in the scale of the Euclidean geometry for the area around the gravitating mass.

Metric of space emitted from a gravitating mass

The local metric is not determined from only the gravitating mass but by the contribution of the gravitating mass integrated with the metric of the large-scale space. The large-scale space can be thought of as a background space.

This background space serves to buffer (lessen) the effect of the gravitating mass, as can be seen as follows:

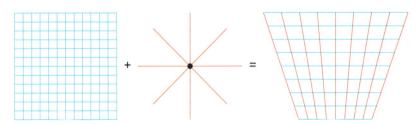

This depicts how the metric changes when the local gravitating metric
is added to the metric of the large-scale structure of the universe.

If Earth were the only matter in the entire universe, then all of space would be emitted outwardly from Earth's center, and there would be no buffering effect from an existing background grid. A falling particle would experience a very significant change in scale as it fell inward toward the center of Earth. It would, in fact, fall at nearly the speed of light. When objects are falling at near the speed of light, it does not make for a very hospitable environment on the surface of Earth. Likewise, a universe made only of the solar system or Milky Way galaxy would also be much different than what we experience with the buffering effect from the entire large-scale structure of the universe. It is the space emitted from a local mass as compared with the space emitted by the entirety of the universe that determines how an apple will fall to the surface of Earth. Notice that we are not curving space or space-time; we are simply changing the scale of the metric as determined by the distance between two adjacent free particles of space. This topic will be further developed in a quantitative manner in chapter 14.

7

SPACE: A THREE-DIMENSIONAL ENTITY

Before we leave the topic of space, there is more to consider. Defining the metric as the distance between two free particles does not provide us with a clear definition of space. The metric is only one-dimensional. The distance between two points lies on a straight segment in Euclidean space. How do we get from there to the idea of three dimensions?

Perhaps this seems trivial. We all intuitively understand that distance can play out along three dimensions and that space is simply the three-dimensional reality of distance along the x, y, and z axes of a Cartesian coordinate system (a system where each of the axes are separated by ninety degrees). So you say, "Let's move on, we have this one handled." But not so fast.

Others may be wondering how I can suggest in the previous chapter that the contribution to the metric from a gravitating mass can be displayed on a flat sheet of paper. Reality, at least as experienced by humans, is three-dimensional, and therefore the space emitted by a gravitating body should be emitted in three dimensions and not simply drawn outward on a flat, two-dimensional surface.

So we have a bit more to consider. Hang on, this gets somewhat abstract—but may be well worth the journey!

The word "dimension" in science has a much broader meaning than simply an orientation in space. A dimension is the entity that you wish to describe and/or measure. For example, the number fifty-two has no dimension associated with it. It is simply a number. Now if I say fifty-two meters, then I have assigned the dimension of distance to the number. Time and mass are also dimensions. There are lots of things that can be measured, such as temperature, density, velocity, energy, pressure, voltage, amperage, and much else. But in their purest sense, all physical dimensions can be reduced to different combinations of mass, time, and distance. These are the three fundamental dimensions.

Why bring this up in a chapter about space? Consider this: Isn't it a little curious that there are three fundamental dimensions and three physical dimensions of awareness in what we call space?

Up until this point, we have only looked at how the metric changes with distance, but the metric also varies in time and mass. We simply have not been looking at that, because we have been able to describe gravity only in terms of the change in the distance metric. This is what is most obvious when we consider the scale of a map: one inch to one mile or one inch to five miles. We could have defined our map equally well in terms of mass or time: one minute to five minutes, one kilogram to ten kilograms, and so on.

Now if space is defined as the phase change from a bound state to a free state, and if this phase change occurs at a regular rate, then one could easily define space as the age of the universe. The bigger the universe (the more space), the older the universe. Its age (time as determined by a continuously running stopwatch) is calculated exactly by the number of free particles in the universe.

SPACE: A THREE-DIMENSIONAL ENTITY

But wait—space as defined by free particles also has mass. So, if I want, I can define space as the total mass of those free particles.

Lastly, I can define space in any combination of units as long as in total it is limited to the three dimensions of physical awareness. Mass, distance, and time are all one-dimensional entities. When combined, they make up three different dimensions: the three dimensions of physical awareness and, perhaps, the three dimensions we think of as space.

You may very likely reject this suggestion. But, if so, why?

When we were young, most of us were taught to think of space as being composed of three dimensions of distance. But this doesn't make sense. Distance is only one-dimensional. Why would the concept of distance ever need to be three-dimensional? Using the same dimension three times doesn't make sense when attempting to define what makes a dimension different. Three different dimensions of the same dimension is actually somewhat absurd. Simply because this is the way we have been taught to intuitively think about space does not mean it is the only way we can think about space. Why can't I define space as time (the age of the universe) or as mass (the total mass of free particles) or as a combination of mass, distance, and time? The answer: I absolutely can. But if you are not quite there yet, that's OK—we will visit this topic again in future chapters. So for now, let's move on.

8

THE ENERGIME

A point is an interesting concept in both daily life and within mathematics.

It is a curious entity and in its most abstract form represents a marker that identifies something without taking up any space of its own. Of course if we wish to see a point, we must actually make a mark with a pencil or pen. The finer the graphite in the pencil, the sharper and more defined the point is. We can combine points to make a segment, yet this is an abstraction because none of the points take up any space. A point does not define any distance, so it is curious how any combination of points can actually define a segment. We often graph things by looking at how the individual points all line up and then try to establish meaning and relationships from a combination of points.

What is my point? There is nothing in nature that suggests a point should have such a significant status. The point is a purely human-made concept and one that we use to provide structure, order, and meaning. We, therefore, tend to extend this human-constructed idea to think that nature must certainly also be comprised of points and that to understand nature we must assign a point to all things important. As an example, in GR every event that has ever occurred in the past fourteen billion years is assigned a point in space-time.

But what if nature is organized differently?

In fact, if you were to build something—let's say the universe—would you simply fill everything up with points? Or might you use something with a little more structure? Something like a line segment.

Let's start smaller. If you were to build a house, would you simply combine points (specks of sawdust) to establish the structure of your house? Or would you use two-by-four studs to define and build the fundamental structure? What if nature has no interest in points but rather builds the universe up with segments?

This is not a new idea. String theory is based on the idea of segments rather than points. But the segments in string theory are on the level of the microworld; they are very tiny segments. Let's think bigger: What if the entire structure of the universe is constructed with segments?

Think of it this way: a forest has many trees, but each of the tree trunks extends upward as a segment. The number of trees can be determined from a two-dimensional surface. You simply need to walk along the ground and count the number of trees. But to determine the volume of the forest, you must take into account the height of the trees. Let's make this simple and say all the trees are equally spaced from each other with the tree trunks extending straight upward with the same height. We can then determine the volume of the forest by simply counting the number of trees on a two-dimensional surface. And the overall structure of the forest is determined by segments (trees) rather than points (saw dust).

From this point forward, these segments of space will be referred to as *energimes*.

If you are interested in the derivation of that word, it comes from my early research when I was exploring the existence of a fundamental energy-time particle; hence, it was shortened to simply energime. We will see in future chapters that the properties of the energime are actually more fundamental than a relationship between energy and time, but the name has stuck with me.

The three-dimensional volume of the forest can be defined
from the number of trees extending upward
from the two-dimensional surface.

9

QUANTIFYING THE ENERGIME

Remember in the first chapter where I stated that to move a scientific idea forward, you need to move beyond words and incorporate mathematics to help quantify what you are proposing? Well, it is time to quantify the properties of the energime.

The energime is a fundamental particle that can exist in a bound or free state. In its free state (space), it defines the mass, radius (distance), and age (time) of the universe. The metric is the property of mass, distance, or time that exists between any two adjacent free energimes. So if we are going to move this concept forward, we must find a way to quantify the distance, mass, and time associated with an energime. If all energimes are identical, we can determine the dimensional properties of a single energime and then determine the large-scale properties associated with the size, mass, and age of the universe simply by knowing the number of free energimes.

Because the energime is so fundamental to our journey, I thought it was important to develop its properties within section 1 of the book. This does get a little heavier on mathematics, physical constants, and scientific equations, but please work your way through this. All the mathematics is clearly explained along the way.

FROM FALLING APPLES TO THE UNIVERSE

The idea of a fundamental value for mass, distance, and time is closely related to what scientists refer to as the Planck scale, named after Max Planck. The general idea is that some of the fundamental constants of nature can be thought of as having a value of unity, which is the same as saying a value of one. The gravitational constant and Planck's constant (used in determining the energy of light at various frequencies) are two of the constants that are assigned the value of one. Planck realized that when you multiply the gravitational constant by Planck's constant you can determine the unity value for distance. When Planck's constant is divided by the gravitational constant you can determine the unity value for mass. By defining the value of the speed of light to be one, you can easily convert the unity distance to a unity time. These unity values can be viewed as the fundamental values of distance, mass, and time. This is all straightforward and has been used by scientists for over a century.

The problem is that it is based on the assumption that the gravitational constant and Planck's constant can be defined to be equal to one. This is solely a human decision and may not be what nature has in mind. It may have nothing to do with reality. It's a nice idea that tends to make some of the mathematics easier. But its entire foundation is built on the idea that we must all agree to accept these fundamental assignments of unity. In reality it might be totally wrong. From a statistical probability, it is most certainly totally wrong.

To make this point, let's take a break for a little story.

10

AN ISOLATED ISLAND ECONOMY

In this story, on an isolated tropical island we find a community of people that has never had contact with any outside currency. One day a ship sinks offshore, and a chest washes ashore that is filled with an equal number of ten-dollar bills and pennies. The isolated community has no way of knowing what the value of the paper money is, but they are fascinated with it and they decide to use it for trading with the idea that each piece of paper is worth one unit of value. This works fine because they all agree that each ten-dollar bill represents a value of one.

After many years, their island economy grows and they run out of paper money from the chest. They decide to use all the pennies, so they again agree to trade the pennies at the same value as the paper money. All works fine in the local economy because all agree to play by the same rules. In fact, the fishermen love the pennies because they don't get soggy when wet. Most of the fishermen trade all their paper money for coins.

Years later the owner of the chest tracks it down. The owner introduces herself to the isolated community and tells them that her name is Nature. As Nature gets to know the people on the island, she realizes that it would be unfair to just demand the money back since it is now part of a flourishing local economy. So Nature decides it would be more advantageous to trade with the island residents, but only based on the true value of the currency. Now the fishermen that exchanged all their paper bills for pennies protest loudly that this is not fair. But Nature says, "I'm sorry, I will trade with you only on the true value of the penny. It is unfortunate that you all agreed to make up a value, but the value you all agreed upon was wrong." You see, in our story, Nature is the ultimate authority, and all others must be subservient to her wishes.

So the fishermen only receive one-hundredth of the value they thought they had when they trade with Nature. Of course no one with paper money is willing to trade with the fishermen on an equal basis once they find out what Nature had to say. The paper holders suggest that the local economy is a constant, so if the fishermen only received one-hundredth of their value, then the paper money must certainly be worth one hundred times what they initially agreed to. So they immediately revalue their paper currency to a value of one hundred to one, thus stabilizing the local economy and benefiting from a tremendous windfall profit at the expense of the penny holders.

Many of the paper holders begin to spend their new wealth very carelessly. But, soon enough, Nature comes knocking at the doors of the paper holders, and they, too, find out that they were wrong. Their paper currency is worth only 10 percent of what they had revalued it to be. Each bill is only worth a unit value of ten, not one hundred as they had thought. While they fare much better than the fishermen, many of the paper holders spent money so recklessly and quickly that they, too, are now in financial ruins. They ask themselves, "How did things get so far out of control? We all agreed on the value of the currency. How did we all end up bankrupt?"

The moral of the story: When we arbitrarily assign values, we make two mistakes. We will almost certainly assign the wrong value to any entity, and we will almost certainly misinterpret the relationship between different entities (one penny is not the same as a ten-dollar bill, and neither is equal to one dollar). Just because we may all initially agree to play by the same rules, that does not make it so.

If we wish to truly understand the fundamental meaning of Planck values, we must understand the nature of unity not as defined by humans—but as defined by the true authority, Nature.

11

THE (INVERSE) FINE-STRUCTURE CONSTANT

In our search to understand the meaning of unity, we have one piece of very lucky information. It turns out that if you combine Planck's constant, Coulomb's constant (used in determining the value of the force between two electrically charged particles), and the value of the fundamental electric charge in just the right way, all the dimensions (units) cancel out and you are left with a number. A number that is not associated with any dimension. Now if the fundamental constants are truly equal to unity (one), then when we combine them mathematically, ideally we should get a value of one. Just use some basic math: one times one equals one, one divided by one equals one, one squared equals one, the square root of one is one. So multiplying, dividing, and determining the square root of any combination of constants with the value of one should give us one.

But what you get is a number very close to 137. What a weird number, right? This idea suggests you should get 1, and you get 137. Doesn't this tell you that your initial premise is probably wrong? You do not need to understand the meaning of 137 to accept the implication that the fundamental constants are most likely not related to unity. Just because it would be convenient if they were related to unity, that does not make it so. We can all pretend they are unity, but we may end up just like the residents of the isolated tropical island. In this case, scientifically bankrupt.

So we need a new approach. Again, I am going to summarize some thoughts in the following paragraph and direct those looking for a more mathematical approach to section 3 of the book.

Scale metrics suggests that Planck's constant (usually denoted with the letter h) defines the radius of the universe. Some readers will be shocked by this statement and wonder why the units for Planck's constant are not then expressed in meters (as opposed to joule-seconds, which are an expression of energy multiplied by time). Using scale metrics, the energy of a photon can be shown to be nothing more than a count of the number of energimes present (a unitless number). This energy is determined from Planck's constant divided by the wavelength of the photon. The units for wavelength are expressed as a distance. Hence Planck's constant can also be easily expressed as a distance.

We will find as we move further into scale metrics that units can become somewhat distracting in that we will ultimately describe the physical constants (along with all measurements) using nothing more than dimensionless numbers representing their true value as defined by nature. In the final analysis, everything comes down to number theory. That is, all you need to know how to do is count and keep track of what you are counting. That should give you a little more comfort if you were worried about being able to handle the mathematics!

Planck's constant represents the radius of the universe when all bound energimes have phase changed to free energimes. It is somewhat related to the concept of the heat death of the universe, a time when all energy has reached an equilibrium and energy is no longer available to do any useful work. Scale metrics further suggests that the gravitational

THE (INVERSE) FINE-STRUCTURE CONSTANT

constant G is inversely related to the circumference of a circle defined by a radius equal to h.

The idea is that h and G are inversely proportional (as one value increases, the other decreases). Using this idea, the fundamental distance associated with the energime is 1.02×10^{-34} meters. From there you can determine the fundamental time to be 3.38×10^{-43} seconds.

If you are interested in the math, it goes like this: We define the speed of light to be one. You say, "How can you do that? Isn't that just making the mistake of the isolated tropical island all over again?" The answer is that to the best of our knowledge, the speed of light is the fastest that anything can be measured to move. It represents 100 percent of the total possible velocity available to any observer. And the numeric value of 100 percent is indeed one. So there is some observational evidence to define the speed of light, which is 2.998×10^8 meters/second, as one. When we do this, all other velocities are then expressed as a fraction of the speed of light. This allows us to express Planck's constant and the gravitational constant as follows:

$$h = 6.626 \times 10^{-34} \frac{kg \cdot m^2}{s} = 2.210 \times 10^{-42}\ kg \cdot m,$$

$$G = 6.674 \times 10^{-11} \frac{m^3}{kg \cdot s^2} = 7.425 \times 10^{-28} \frac{m}{kg}.$$

Where mass is expressed in kilograms (kg), distance in meters (m), and time in seconds (s).

Therefore,

$$1 = \sqrt{h \times 2\pi G} = 1.015 \times 10^{-34}\ m.$$

And dividing this by the speed of light gives

$$1 = 3.387 \times 10^{-43} \text{s}$$

This gives us the unity values for distance and time. These two values are very close to what the current notions of Planck distance and Planck time are. In scale metrics, they are the metric values of the distance and time represented by the interval between two adjacent free energimes. It may be important to note once again that these are an actual physical distance and an actual time. This is different from GR, where the metric is a complete abstraction. The GR metric is an imaginary concept of the distance (or time) between two entities that are not really present.

Now on to the topic of the unity value for mass.

The current idea of Planck mass is that it has a value of 2.18×10^{-8} kg. While this is a small number, it is much larger than other known entities, such as the mass of the electron (9.11×10^{-31} kg) or the mass of the proton (1.67×10^{-27} kg). For many, it raises the question of how this might be considered a fundamental mass (a building-block mass) if there are other masses even smaller than this. Others try and explain this away using philosophical arguments that the amount of mass is a relative concept and that large and small have no inherent individual meaning. There is yet another line of reasoning stating that the focus should not be on why the Planck mass seems so large but why the gravitational influence is so weak as compared with the other fundamental forces.

But here is the problem: if h really does represent the heat-death radius of the universe (which is only an idea), then it must be a fairly large dimensionless number. Remember, I can determine the diameter of the universe by counting the number of free energimes in any one direction of the universe. Just based on the estimated age of the universe,

THE (INVERSE) FINE-STRUCTURE CONSTANT

roughly fourteen billion years, the value of h would be at least 1×10^{60}—that is a one with sixty zeros after it.

If G is the inverse of the circumference, then it is a very small number, only a tiny fraction of one. So when I multiply these two numbers (which is what you do to get the Planck distance and time), it tends to cancel out the differences. Look at it this way: one times one equals one, but one hundred times one-hundredth is also equal to one. In this case, treating G and h as unity gives you the same value as if G and h were vastly different but inversely proportional— that is, as one number increases, the other proportionally decreases.

Now to determine the Planck mass, you must divide the value of h by G. This is going to result in a much different, much smaller value for the fundamental mass than that of the current Planck mass. This tells us why the current Planck mass is way too large, but it doesn't tell us enough to actually determine the value of the fundamental mass. For that, we need some help from another force.

12

ELECTROSTATIC FORCE

Yes, this is a book about gravity, but it may be helpful to understand other forces of nature in our search to understand gravity. Electrostatic force is a good place to start because it is mathematically expressed in a way very similar to that of gravity:

$$F = \frac{kq_1q_2}{r^2}.$$

This is an equation introduced by Charles de Coulomb, with F representing force, k being Coulomb's constant, and the q's representing the electric charge of two particles separated by a distance r. As you can see, the form of this equation is very similar to Newton's equation for gravitation:

$$F = \frac{GMm}{r^2}.$$

Since we are reasonably sure that the fundamental constants are not equal to one, perhaps the value of these constants is associated with the strength of the force they exert. For example, when we compare electrostatic force (the force between two electrons) using Coulomb's equation from above, we find out that the electrostatic force is

4.17 × 10⁴² times stronger than the gravitational force (from Newton's equation) between those same two electrons at the same distance from each other. Would it not make intuitive sense that the value for their constants is related to this difference in the force?

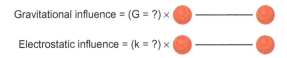

Two pairs of electrons (as pictured above) at the same distance apart display both a gravitational and electrostatic influence. Since the electrons are the same, what then determines the value of the influence may very well be the value of the constants, *G* and *k*.

We had an idea that *G* and *k* were equal to each other and equal to unity, but we were able to pretty easily show that this idea does not really make sense. Now this is a different idea, and the test for it is whether we can learn something from it. In this model we state that *k* is 4.17×10^{42} times larger than *G* because it exerts a force that is 4.17×10^{42} times stronger. We still do not have a specific value for *G* or *k*, but we do have a value for the ratio *k/G*, and that might be enough to move us forward. But first I ask the question: Doesn't this look a lot like the path we just went down regarding gravitation? In this case, Coulomb, as opposed to Newton, suggests that the electrostatic force is proportional to the charge of two electrons multiplied together and divided by the distance of separation squared. Is it possible that this force could also be explained by a changing metric?

Let's think about this. In scale metrics, the reason an apple falls to the earth is because of a changing metric that is determined by integrating the pattern of free energimes

ELECTROSTATIC FORCE

emitted from the gravitating body with the background space representing the free energimes emitted by the entire large-scale structure of the universe.

The large-scale structure of the universe would not be expected to change simply because we wish to introduce a new type of force. So if the electrostatic force is 4.17×10^{42} times stronger, then there must be 4.17×10^{42} times more free energimes emitted in the local pattern of space by an electron propagating electrostatic force than an electron propagating gravity. Remember that the electron does both. It has mass, so it has a gravitational influence on the metric, and the same entity (the electron) emits an electrostatic influence. Yet the electron is an electron. It would not make any sense to say that when it produces gravity the electron is one thing and when it emits an electrostatic influence the electron is something else. You might be able to get away with this in an abstract mathematical sense, but from the perspective of observational science, an electron is an electron regardless of what it is doing.

Let's take this information and see if we can learn anything from it.

We can use the electrostatic force equation to determine that the force between two electrons separated by one meter is 2.31×10^{-28} N. I have suggested that k is 4.17×10^{42} times larger than G. Therefore, we can substitute $4.17 \times 10^{42} \cdot G$ for k. We can now solve for the charge of the electron in units of kilograms. What do we get? The charge of the electron is 9.11×10^{-31} kg. This is exactly the same as the mass of the electron. This seems really exciting, except in full disclosure, once we set the two constants (k and G) to have a ratio of 4.17×10^{42}, we basically set in motion that the charge of the electron would be equal to the electron mass. So this is a good result but nothing to be really excited over.

There is still more to learn, though. If the charge is 4.17×10^{42} times stronger than gravity, then the entity responsible for propagating gravity (the energime) must be 4.17×10^{42} times smaller than the electron, or 2.19×10^{-73} kg. We finally have a fundamental mass of the energime. Now we should be excited!

In case that just happened too fast, let's break it down a little more.

Think of charge just like a shotgun shell. You can shoot BBs one at a time from a BB gun all around in a circular pattern. Or you can shoot an array of BBs from a shotgun with each shell filled with a charge of BBs.

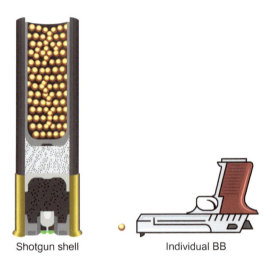

Shotgun shell   Individual BB

Electric charge is perhaps the same concept. Gravity is from single energimes being emitted from the electron. Electrostatic force is from a charge (grouping of energimes) being emitted. This charge (container) holds 4.17×10^{42} individual energimes and manifests itself as the fundamental electric charge.

The electron charge is equal to the mass of the electron and is composed of 4.17×10^{42} energimes. Hence each energime has a mass of 2.19×10^{-73} kg. Notice that the fundamental mass of the energime is now many, many factors of ten smaller than the mass of the electron, proton, or any other known particle. It truly becomes a building-block mass from which all other masses can be constructed from.

We have now completed the goal of establishing the unity values for mass, distance, and time!

Scale metrics suggests that each free energime defines space as three-dimensional and is composed of a combination of the one-dimensional values of distance, time, and mass with specific metric values of

$$1.02 \times 10^{-34} \text{ m,}$$

$$3.39 \times 10^{-43} \text{ s,}$$

$$2.19 \times 10^{-73} \text{ kg.}$$

We are now also able to determine the unitless values of G and h. Planck's constant, h, has a value of 9.95×10^{64}, and the gravitational constant (G) has a value of 1.60×10^{-66}. Notice that neither of these numbers has any units associated with them. These are the true fundamental values of G and h. As we thought, G and h have very different values that become obvious when you divide the two numbers. But when multiplied, they behave just as they would if they were both equal to one along with a proportionality constant of 2π that relates the radius of h to the circumference of $1/G$.

Before we leave this chapter, some may ask why I used the mass of the electron as opposed to the mass of the proton to determine the mass of the energime. This is an excellent question. Both carry the same charge, so it might seem arbitrary that I chose one over the other. And if I had used the proton, that would have given me a different value for the mass of the energime. So how confident can we be in this fundamental energime mass?

Several things to consider: The electron is recognized as a fundamental particle and is also the smallest known mass that carries a full electric charge. The proton, on the other hand, is not a fundamental particle and in addition to propagating an electrostatic force also plays a role in the strong nuclear force. Therefore I appropriately chose the electron as the fundamental electric charge in the calculation to establish the energime mass.

One further point: Using the energime mass as determined from the charge of the electron, we can predict the electron mass, proton mass, and neutron mass. We can also develop a model of the electrostatic force and the strong nuclear force and unify these with the notion of a quantum model of gravitation. These outcomes provide fairly strong evidence that we may well be on the right track. These topics are covered in chapters 36–39.

13

A QUICK REVIEW

Let me tie some of the main points of previous chapters all together. I have suggested that space is created from the phase change of an energime from its bound state to its free state. The metric associated with this is simply the distance between any two adjacent free energimes. These energimes are real, and therefore the distance between them is real, not just an abstraction, as is the case for GR.

However, we do not need to limit the metric to a distance. Space can be viewed as a combination of the energime's mass, distance, and time. Each of these one-dimensional entities combines to form three dimensions of physical awareness. This awareness is what we refer to as space. The values of the energime can be quantitatively determined as

$$\text{Energime mass: } 2.19 \times 10^{-73} \text{ kg}$$
$$\text{Energime distance: } 1.02 \times 10^{-34} \text{ m}$$
$$\text{Energime time: } 3.39 \times 10^{-43} \text{ s}$$

Each of these values represents a true unity value as defined by nature (as opposed to humankind). Therefore you can express them as all being equal to unity:

$$1 = 2.19 \times 10^{-73} \text{ kg} = 1.02 \times 10^{-34} \text{ m} = 3.39 \times 10^{-43} \text{ s}$$

However, this does not make mass, distance, or time equal to each other. Rather, these are three different values of unity, but they all represent something quite unique. Therefore they are not interchangeable. In scale metrics we would never express time as having distance-like properties nor can distance ever be considered time-like. This is why they are all expressed as different dimensions that are orthogonal to each other on the three axes of a Cartesian coordinate system. Each energime is one particle manifested, in every instance, as three unique dimensions.

The large-scale structure of the universe provides a background space (or grid), and a local mass (such as Earth) provides a pattern of space emitted from its center outward that must be overlaid onto the background space. This provides a changing metric (just as the scale of a map can change) that we can use to determine how an apple will fall to Earth. This is accomplished using the Pythagorean theorem, with the hypotenuse being the value of the metric at infinity, one leg representing the value of the local metric, and the other leg providing the velocity squared of a falling object originating at infinity and falling to a specific distance r. This in many ways is very similar to GR. The primary difference is that scale metrics uses a changing scale to vary the metric, while GR uses a curvature of space-time.

We know that in GR, plotting a time dimension serves to buffer the curvature of space-time when time is multiplied by the speed of light to give us a value of distance. This time-related distance (which is not allowed in scale metrics) is vastly longer than any physical change in distance in most real-life situations, resulting in a space-time curvature that is very nearly a straight line.

When using scale metrics, it is the background metric from the large-scale structure of the universe that buffers the change in the scale of the metric. We know that the size of the universe is equivalent to a measure of time. In this way, both GR and scale metrics are chasing the same basic concept in attempting to model how time impacts the value of the metric. However, there are subtle differences in the predictions made by these models. And it is these differences that can help differentiate between the models and determine how well they mimic and simulate reality.

14

A QUANTITATIVE LOOK AT GR VERSUS SCALE METRICS

In GR, when viewing a time-static snapshot of a cross section of space-time, we see that space-time bends like a rubber sheet with a weight on it. Anyone with the most basic interest in GR has probably seen this example.

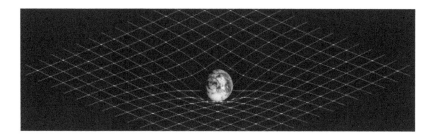

This is an often-used analogy that, in my opinion, is a poor one. First, it is strange that the rubber-sheet analogy attempts to explain the space-time-warping property associated with gravity but must use gravity in the process. Gravity is what pulls the weighted ball downward and warps the rubber sheet. A model that uses gravity to explain gravity is significantly limited in its usefulness. Second, and perhaps fundamentally more important, the analogy suggests that a test marble rolling around and eventually downward is somehow falling toward the gravitating mass. In reality, the image should be rotated 90 degrees because what is shown on the diagram to be the horizontal path is actually the downward "falling" path. By rotating the picture 90 degrees and allowing a test particle to fall directly to the surface of Earth, it better illustrates the test particle actually falling downward (as defined by coordinate distance) within an overall geometry that is indeed curved (defining the proper distance).

Notice that because it is curving, the proper distance (blue) is somewhat longer than the coordinate distance (red).

In GR, the mathematics for determining the ratio of a tiny interval of proper distance, ds, to the corresponding tiny interval of coordinate distance, dr, at any distance r can be expressed as

$$\frac{ds}{dr} = \frac{1}{\sqrt{1 - \frac{2GM}{r}}}.$$

Here, *G* is the gravitational constant, *M* is the mass of the gravitating body (Earth), and *r* is the distance of separation between the center of Earth and the test particle (an apple) falling toward the surface of Earth.

Within scale metrics, we see a much simpler picture (shown below). As the apple falls directly to the surface of Earth, it passed through maps that are defined by different scales.

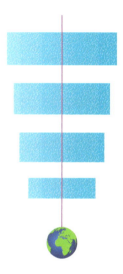

In my mind, this is a much easier picture to process—no curved space-time, simply a change in scale for the Euclidean space of each map you pass through as you fall toward Earth. We also have a clear understanding as to why the scale changes. Remember, the space pattern of the gravitating mass must be integrated into the space grid of the large-scale structure of the universe, resulting in a change in scale at different locations. In other words, we incorporate the general notion of Mach's principle into our understanding of gravitation, in that we allow the large-scale structure of the universe to have an impact on what we observe locally, with all the geometries neatly defined by Euclidean space. As the apple falls to Earth,

FROM FALLING APPLES TO THE UNIVERSE

the value of the metric (scale of the map) becomes smaller. If we define the local metric as b, relative to a metric value of one for a location far, far away, then b can be expressed as

$$b = \frac{1}{1 + \dfrac{GM}{r}}.$$

See section 3 for a derivation of this equation.

Remember that models are developed to mimic reality. We have two different models that use different equations to quantitatively define their metric. Both models use the same concept of a metric but determine the value of the metric differently. It should be pretty easy to see which one works better, right? This is the interesting thing about models and mathematics. Within small limits—and a tiny range of values—different mathematical equations can provide the same results. Within all observable experiments (phenomena that has actually been observed as opposed to outcomes that are influenced by models), there is no difference between the two methods suggested by GR and scale metrics.

In other words, expressed mathematically, we can say that for all directly observed data

$$\frac{1}{\sqrt{1 - \dfrac{2GM}{r}}} \approx 1 + \frac{GM}{r}.$$

Note: In chapter 21 we will modify the GR expression above to establish an absolute equality between GR and scale metrics.

The curvy equal sign indicates that something is approximately true. For the case above, both sides are unmeasurably close in value. This is because gravitation is such a weak force that it is nearly impossible to get different results from either of the above expressions. It takes some people by surprise when they hear that gravity is a weak force. Many think of it as strong because it hurts a lot when you fall and hit the ground from even a short distance. But think of it this way— Earth is overwhelmingly large as compared with the mass of an electron. It takes a lot of mass to generate all the gravitational influence of Earth. On an equal basis of charge versus mass, the electrostatic force wins out big over the influence of gravity. So, yes, gravity is a weak force.

As to finding no difference between the two expressions, try it yourself if you like. For a test particle (an object small enough that its mass is insignificant as compared with the mass of Earth) falling from infinity to the surface of Earth, the value of GM/r is equal to 0.000000000695983, or 6.95983×10^{-10}. This is the same value used in both GR and scale metrics.

The GR equation provides the following value of ds/dr: 1.000000000695983.
The scale metrics equation provides a value of $1/b$: 1.000000000695983.

These numbers are exactly the same within the limits of any known instrumentation. The fact that the ratio of the metric values is so close to one is further evidence of just how weak gravitation is.

But there is more to consider. In GR you are constantly taking something away in the denominator (bottom part) of the equation. In scale metrics you are constantly adding something into the value of the numerator (top part of the fraction). As just shown above, for very weak gravitational influences, these two approaches can be used to provide the same result. However, at theoretical levels, in the GR model the denominator will eventually go to zero. That is, if I keep taking something away, eventually I have nothing left. If you remember your high school algebra, then you know that zero in the denominator is a nasty situation, leaving you with a value that is either undefined or that goes to infinity, depending on your preference. In physics talk, you simply say the equation blows up!

Note that this never happens in scale metrics. You can continue to add any number you want all the way to infinity and the equation never blows up! It simply drives the value of b to zero, which is perfectly allowed.

Who cares about these theoretical limits? Well, these are the conditions that establish the event horizon of a black hole. And since black holes are a hot topic in science (and science fiction), it might be important to understand what we can learn from various models of gravity.

15

KINETIC ENERGY

Before we tackle black holes, we have some more work to do. For starters, we need to better understand the relationship between energy and velocity for a particle moving within a gravitational influence. When an apple falls to the ground, it gains kinetic energy. This is the energy an object acquires from its motion.

You may remember the classical equation for kinetic energy as being

$$E = \frac{1}{2}mv^2,$$

where m is the mass of the moving object, v is its velocity, and E is the object's kinetic energy.

This classical expression of kinetic energy is often referred to as part of Newtonian mechanics; however, it is not accurate to state that this was one of Newton's direct contributions. Any number of scientists and philosophers over the centuries have suggested a relationship between the mass of an object and its velocity squared. In fact, the use of the term "energy," at least as used by scientists today, was not even introduced until the 1800s by Thomas Young, well after the death of Newton. However, because of Newton's significant contribution in using mathematics to quantify physical observation, the whole category of the classical physics of motion is often simply referred to as Newtonian mechanics.

What do we mean by a classical equation? For our purposes, it means that the mass of a moving object is not affected by the amount of energy it contains. You can see how this was easily assumed to be the case centuries ago. If you think of an object as having an inherent mass to it, then when you input energy, that fixed amount of mass begins to move. In the classical interpretation, there is no direct relationship between the kinetic energy of an object and the object's mass. Adding energy has no impact on the mass of the object.

This interpretation changes when you consider relativistic physics, where it is understood that both the energy and rest mass (the mass of the object when it is motionless) contribute to the overall energy content of the object. The term "energy content" is used to describe the total mass of an object, both its inherent rest mass and the mass it possesses due to the energy it contains. To put it in simpler terms, the mass of an object in classical mechanics stays the same regardless of how fast the object is moving. In relativistic physics—since energy has mass—the faster an object is moving, the more energy content it has, resulting in an ever-increasing relativistic mass.

If the energy content did not increase, there would be no upper limit to the velocity of an object as you added more energy. This would be inconsistent with all known observations that limit the observed velocity of an object to be the speed of light.

As is the case with most classical equations, they tend to work just fine within the limits of our normal experiences. It is only in extreme conditions when classical equations begin to fail. If we remember that an equation is merely a quantitative expression of some model, then we can think of classical models as ones that mimic reality fairly well under normal conditions. But they are not absolute, and they break down quickly when extreme conditions of very high energies and very fast velocities are reached.

So if the classical equation is not correct, is there a better model—or a better equation—to determine kinetic energy for both normal and extreme conditions?

Generally, the relativistic treatment to determine kinetic energy uses what is called a Taylor series. This is the expansion of a function that sums an infinite number of terms to achieve a finite value. It provides a mathematical way to accurately approximate the kinetic energy of a moving object by adding together the first several terms in the series. However, to determine the correct value would theoretically require adding an infinite number of terms together. Therefore, in any practical application, a Taylor series gives us only an approximation, but it is a darn good one.

We are not going to discuss any further details on a Taylor series, because there is a better way to determine the exact kinetic energy of an object traveling at any velocity.

To create this equation, you must make several observations.

At low speeds, the classical equation $E = 1/2\ mv^2$ works quite well. But we also notice that as an object approaches the speed of light, the classical equation does not work. Under these conditions, the correct equation becomes very close to Einstein's famous equation $E = mc^2$.

In Einstein's equation, E once again represents energy, the m represents the mass of the total energy content of the object (its rest mass and the mass associated with its energy), and c stands for the speed of light. When the velocity of an object moves closer to c, its mass m will become increasingly large as the velocity approaches the speed of light.

But also notice that the 1/2 has disappeared in the Einstein equation. Or perhaps a better way to state this is that the 1/2 from the classical equation has become a 1 in the Einstein equation:

$$E = 1/2\, mv^2,$$

$$E = 1\, mc^2.$$

Also note that both v and c represent velocities. We see that these two equations are similar in many ways but represent two extremes. On the one hand, the classical equation works best if the particle is barely moving. That is, if its velocity is approaching zero. The Einstein equation represents an object that is approaching the maximum velocity of the speed of light.

What happens between these two extremes? Well, we are going to change the constant (1/2) to a variable, x. The allowable values for x range from 1/2 at slow speeds to approaching 1 at nearly the speed of light. If we can determine how x changes with velocity for values between zero and c, we will have a better understanding of kinetic energy.

In addition, we must understand how the mass represented by the total energy content changes with the velocity of the object. We could achieve this if there were some factor that we could multiply by the rest mass to get the total energy content of an object at any velocity. We will call this variable gamma, as represented by the Greek letter γ. If we can determine the values for x and γ, we will then have an exact mathematical expression for the kinetic energy of an object moving at any velocity.

Let's start with γ. This has actually been understood for well over a century. The Lorentz transformation equation can be used to determine the exact value of γ. Mathematically, this is expressed as

KINETIC ENERGY

$$\gamma = \frac{1}{\sqrt{1-\left(\dfrac{v}{c}\right)^2}},$$

where v is the velocity of the object and c is the speed of light. As you see, γ changes with the velocity of the object. Multiplying γ by the rest mass of the object (γm) provides the mass of the total energy content of an object moving at velocity v.

Now let's tackle the x variable. This one is not as well documented but can be easily determined. You can think of x as if it represents the orientation of the energimes within an object. For example, you can think of an object at rest as if it were instantaneously composed of energimes all moving outward from the center.

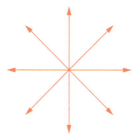

An object at rest

These arrows are actually known in mathematics as vectors, and you can add them all together by placing the beginning of one vector to the end of the next. Picture this in your mind: if you added these vectors, since they are all going outward from the center, they would cancel each other out, representing an object that is not moving—in other words, an object at rest. Some readers might question this and say the object looks like it is exploding. With time, all the energimes are flying away from each other. How is this at rest?

Great question. But what we are looking at is a snapshot. There has been no passage of time; we are simply drawing a representation that is frozen in a single instant.

To make the object move, you could add more vectors that are oriented only along one direction (represented below by the blue lines). When you do this, only half the vectors for the object at rest (represented by the red lines) are oriented along the line of motion, while all the blue vectors are moving in the line of motion. The blue vectors are also not parallel to each other (at least at slow speeds).

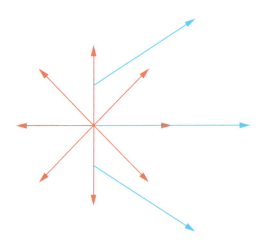

An object with little motion

If I had nearly pure energy (shown with the blue vectors below) with very little rest mass, all the vectors would be aligned in one direction along the line of motion. It turns out that the faster the object is moving, the more energy (blue vectors) are added, and these vectors become more horizontally aligned. As the motion increases, the blue vectors become more parallel. At the speed of light, they will actually achieve a parallel orientation.

KINETIC ENERGY

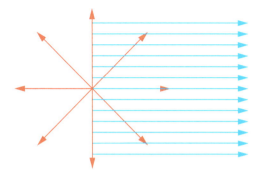

An object with motion approaching c. The energy vectors align in the direction of motion and become increasingly parallel.

If the object is at rest, half the energimes are moving in one linear direction and half are moving in the opposite linear direction.

So $x = 1/2$ represents an object at rest (all red), and $x = 1$ represents energy moving at the speed of light (all blue). The value of x is determined from the combination of both blue and red vectors and defines the overall orientation of the object.

Again, this is a snapshot in time. This is a very important distinction that becomes critical in future chapters. Scale metrics treats motion as a static property of matter. That is, the object does not need to change its position to possess the property of motion. Motion is simply a diagram in static time that shows the pattern of energimes and is quantitatively defined by the variable x.

From here it can be shown that for motion between zero and c, x can be determined as

$$x = \frac{M_o + M_e}{2M_o + M_e},$$

FROM FALLING APPLES TO THE UNIVERSE

where M_o represents the rest mass of the object and M_e represents the mass associated with the added energy.

To better understand this equation in an intuitive way, let's take a quick break for a story about moving day. For those looking for a more rigorous explanation, please refer to section 3.

In this story, I have items of various sizes that need to be moved. While I am not sure how all these items will be packed into boxes, I do know the total weight of everything to be moved. Because of the different sizes, some objects will fit in three-foot boxes and some will need six-foot boxes (the height and depth of each box is one foot). Each of the boxes can handle the same weight limit, and I will make sure each box is packed to its limit. I do not know how many three-foot or six-foot boxes I will need, but I am confident that I will not need more than six boxes altogether. Since I do not have much to move, I am going to share a moving truck with a friend who will use the whole front part of the truck, leaving me with only the back end for my items. I remind my friend to rent a truck with an opening that is at least six feet wide and six feet high. This way, I know I will be able to fit any combination of three-foot and six-foot boxes.

On moving day, it turns out that I used only one six-foot box and five three-foot boxes. No problem, they all fit in the truck with room to spare. But if I want to determine how efficiently I utilized the space in regard to how much weight I could have packed, then I can do this by considering the following illustration.

It turns out that I had room in the truck to fit five additional boxes. So how well did I do?

I packed six boxes but could have packed eleven. All the boxes have the same weight, so I can determine how

KINETIC ENERGY

There is room for five additional three-foot boxes.

efficient I was by taking the ratio 6:11, or 0.55. Now treat the three-foot boxes as being packed with the mass of an object at rest, and treat the six-foot box as the mass of the added energy. How do I find the value of x? I take all the mass I have ($M_o + M_e$) and I divide that by all the mass that could have fit ($M_o + M_e + M_o$) or ($2M_o + M_e$). Notice that if I have all three-foot boxes (object at rest), then $x = 1/2$. If I have all six-foot boxes (added energy), then $x = 1$. For all situations in between, I can use the equation above to determine the exact value of x.

So now we have an exact equation for determining the kinetic energy of an object at any velocity:

$$E = x\gamma mv^2.$$

This little equation will prove to become very valuable as we move forward. It is so much simpler than a Taylor series and actually more accurate. In fact, once you discover that the kinetic energy associated with an object at any velocity can be accurately expressed using the above equation, you immediately realize that the classical equation should be abandoned.

Expressed mathematically, the classical equation should really be stated as

$$E \approx 1/2\,mv^2.$$

The classical equation for kinetic energy does not represent a true equality.

If you want an absolute equality, you must use this version:

$$E = x\gamma mv^2.$$

This is because the classical equation was developed before scientists were aware that adding energy changes the relativistic mass of an object. Therefore the only time the classical equation for the energy of motion is technically correct and provides a true equality is when there is absolutely no motion at all. Yes, that is ironic.

Once motion begins, the value of x must be greater than 1/2 and the value of γ must be greater than 1. This is an extremely important idea. We will find in upcoming chapters that replacing the term 1/2 with our updated notation xy brings GR into complete agreement with scale metrics and has far-reaching implications for black holes! These topics are taken up in chapters 21 and 22.

16

THE ENERGIME: IS IT THE SMALLEST PHOTON?

As we continue to move closer to our goal to address black holes, the question arises as to whether the energime is actually a form of light (electromagnetic radiation). If a pulse of light is composed of energimes all moving in a uniform direction as pure energy, then might it be possible that the energime is actually the smallest possible photon? (That is, the smallest possible quantity, or quantum, of light.)

It might be tempting to think of the energime this way, but it would be incorrect. While there are similarities, there are also significant differences between the energime and a photon. For one, a photon is affected by a gravitational influence, whereas the energime is not. The pattern of free energimes (that is, the local pattern from a gravitating mass overlaid onto the large-scale structure of the universe) determines the gravitational influence. The distribution of free energimes defines gravitation, yet the energimes themselves are not subject to the influence of gravitation.

The energime is not influenced by a changing metric and perhaps even exists in a different metric from the one that defines the speed of light. This becomes tricky and may require that you alter your ideas around the speed of light a little. You have probably heard over and over again that the speed of light is a constant—that it always moves at the same speed.

This is not entirely correct, as any physicist can attest. The correct statement is that any observer will measure the speed of light to be constant within their local surroundings and from that observer's unique orientation within space. How is this different?

We already know that there is often a difference between what one sees locally and what is happening globally. You perceive the floor beneath you to be flat, yet you are on the surface of a sphere. So every observer within any local frame of reference measures light to be the same speed. But that is not the same as saying the speed of light is the same everywhere. What does this mean for scale metrics?

We know that in scale metrics, the metric (the scale of our local map) changes as you move to different positions relative to a gravitating mass. Each observer, from the perspective of that observer's local map, sees the speed of light to be constant. But you can easily see that between the different maps, as the scale changes, the speed of light changes.

This begins to paint a picture that will become clearer in future chapters. It suggests that we will discover a difference between what we see locally and the way the universe really exists. It turns out that light travels within a local metric, as defined by our proper scale. The energime's speed is determined by a global scale that was defined at the earliest moment of the universe, before the flow of time began.

Graphically, it can be displayed as shown on the next page.

This is a key piece of information that will be critical in helping us understand the physics of motion and more specifically the concept of momentum! Let's explore why.

THE ENERGIME: IS IT THE SMALLEST PHOTON?

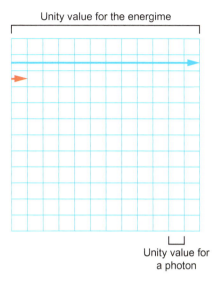

The energime motion vector (blue) is defined by the global metric defining the universe. The motion vector for light (red) is defined by the local metric.

17

MOMENTUM

We have all heard the term "momentum" and have probably used it in everyday situations. The idea of momentum is that once we start moving, we tend to keep moving. If you are working on a project and have great momentum, it is going to be hard to stop you. The football team with momentum is likely to march right down the field and score. In physics, the equation for momentum is mass times velocity, mv. The more mass something has and the faster it is moving, the harder it will be to stop its forward progress. If you were to position yourself in front of an approaching object that you were going to have to stop, you would be hoping it was not too massive or moving too fast. From this regard, momentum is intuitive. But that is not always the case, particularly when we try to understand momentum using only classical mechanics.

More advanced readers might state that I need to approach this subject using four-dimensional space-time. But my goal is to explain physical observations within our three dimensions of awareness by utilizing Euclidean geometry. So let's explore momentum from this perspective. For starters, momentum is always conserved in any interaction. This means if you add all the momentum for a group of particles before a collision and then add the momentum of the same system of particles after they collide, the total momentum will be the same.

Let's just look at a system of two identical objects. The first is moving toward the second, which is at rest. After they collide, they are going to stick together and make one larger (more massive) object. The momentum of the first object is mv, and the second is at rest, so its momentum is zero. This means the total momentum before the collision is mv. After they collide, the mass of the combined objects is twice as much as the original moving object, $2m$. Since the momentum must be the same after the collision, we can easily see that the velocity of the combined particles after they collide will be $v/2$.

What we see happening above is that the final velocity of the combined objects can be determined by taking the total momentum that was absorbed and dividing it by the total energy content of the objects. Expressed as an equation:

$$\frac{\text{Absorbed Momentum}}{\text{Total Energy Content}} = \frac{mv}{2m} = \frac{1}{2}v.$$

Note that the above equation is a classical treatment of the collision, because both the absorbed momentum and total energy content are determined only by the mass of the objects without accounting for the mass of the energy required to provide motion to the objects. And, as a model, this works fine because most of the absorbed momentum comes entirely from the rest mass of the object.

This is fairly straightforward, but it does not allow you to make up any random collision even if you ensure that the momentum is conserved. It turns out that nature has some rules that govern the outcome of collisions. These rules are quite simple. You need to account for both the conservation of momentum and energy. When you do this, there are two possible situations:

MOMENTUM

- In one case, the momentum is conserved, but some energy seems to have been lost after the collision. It turns out that the energy is not really missing; rather, it simply leaked out of the system, usually in the form of heat. This is called an inelastic collision. It generally occurs when the particles stick together or are dented or altered in shape because of the collision. You can imagine that the friction created by the particles rubbing and pressing together, along with the energy that alters the shape of the objects, pulls away some of the energy that otherwise would have been used to move the particles. Our example from above was for an inelastic collision.
- In the second case, the momentum and kinetic energy are both conserved. This is called an elastic collision. This generally occurs when all the objects bounce cleanly off each other with no damage (dents, dings, etc.). In this situation, no significant friction occurs between the objects, and all energy is applied toward the kinetic energy of the particles.

So let's go back to our two objects and look at an elastic collision. In this case, let's use two balls that will cleanly bounce off each other. In this case, the balls will simply switch roles (assuming they hit head-on and have the same mass). The ball that was moving becomes stationary, and the ball that was stationary is now moving at velocity v, which was the velocity of the first ball.

The above is a nice classical explanation, but now let's think a little deeper and incorporate what we have learned about energy content. Wasn't energy transferred when the balls collided? And doesn't all energy have a mass equivalence? So shouldn't that transferred energy contain momentum? The answer to all three of these questions is a resounding yes. So let's think this through to its logical conclusion. From what we know, we should be able to determine the velocity of the object by dividing the momentum absorbed during the collision by the total energy content of

FROM FALLING APPLES TO THE UNIVERSE

the object. After all, that is exactly what happened when one particle absorbed the other in our first example of an inelastic collision. The only difference is that the momentum of the absorbed object was so large that we could ignore any increased mass caused by the total energy content.

In the case of the elastic collision between two balls, the only absorbed energy was from the increased energy content of the ball (as opposed to both the energy and mass of the entire ball), and here we cannot ignore it. We have energy being absorbed by the object, and this energy has mass. We can therefore state that it also has momentum.

Why shouldn't we expect to be able to determine the object's velocity by dividing the absorbed momentum by the total energy content? Excellent question. And if you do this calculation, you will find that it does not agree with observation. The above thought experiment does not work. What!

I like to think of it this way: Let's say you have two trucks loaded with gravel heaped in a pile high above the sidewalls of the truck bed. The first truck collides with another identical gravel truck. When the one truck hits the other, the first truck stops due to the impact, and the second truck lurches forward. With such a sudden stop, some of the gravel from the first truck flies across and lands in the bed of the second truck. Think of the gravel that flew across as the transferred energy. The first truck is no longer moving, so it lost some of its energy content. The second truck is now moving forward and has gained energy content exactly equal to what the first truck lost (our example is for an elastic collision). And why did the gravel fly between the trucks? Because its momentum kept it moving forward. Doesn't that sound like a fairly straightforward analogy? It works perfectly to address the change in energy content of the trucks before and after the collision. But for some reason, it does not work for momentum. The mass associated with the transferred energy (even if moving at the maximum speed of light) does not provide enough momentum to account for the velocity of the second truck. Why would this be? I followed the rules. In an elastic collision, both momentum and kinetic energy should be conserved. Mass was transferred in the form of energy, and that mass should also possess momentum.

Here is a secret on classical mechanics: Scientists talk all the time about transferred energy. But you will not often find a reference to transferred momentum. The momentum before and after the collision is conserved—that is true. But this conserved momentum is achieved without considering momentum as being transferred between the objects. It appears as if the object receiving energy gains far more momentum than can be expected from any transfer of momentum. And the object losing energy also loses far more momentum than can be expected. The final result—momentum was conserved. But if you try to figure out the detail in the middle of that interaction—that is, the momentum that was transferred from one object to the other—you will get the wrong answer. Apparently you cannot visualize the transfer of momentum as you can with the transfer of energy.

To me, that has always seemed absurd. We know that energy has mass. If energy is transferred, then mass is transferred. If this transferred mass has a velocity, then it has a momentum. Having realized that, how can you avoid the concept of transferred momentum?

We have already developed the equation for this:

$$\frac{\text{Absorbed Momentum}}{\text{Total Energy Content}} = \frac{m_e c}{m_o + m_e}.$$

Yet the observed velocity of this interaction is many times larger than what the above equation suggests. What is going on?

A possible solution can be found in scale metrics.

Remember from the previous chapter that the energime operates within a different scale (metric) from the local metric of the object.

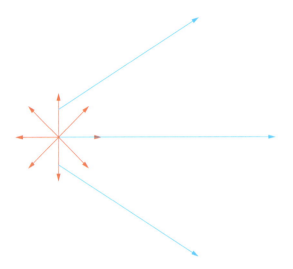

The blue energy vectors reside in the much larger metric defined by the global scale of the universe. Their orientation to the direction of motion is determined by the value of $\sqrt{x}$.

The transferred energimes have a motion vector equal to c as defined by the global scale of the universe. The object absorbing this energy resides in the local metric (local map). The result is that the transferred energimes can be viewed as having a motion many times the speed of light as related to the local metric of the object.

This makes a huge difference and provides a good way to test how well scale metrics mimics actual observation. Scale metrics predicts the need for a scaler (a number we can multiply by the momentum of transferred energy) that accounts for the difference in the metric of the energime as compared with the local metric. Put another way, it provides a ratio of

MOMENTUM

the energime's motion vector to the motion vector of light. While we discuss this in more detail in section 3, there is good reason to expect the scaler to be related to the inverse of the square root of the momentum derived value for velocity (plus an adjustment for the orientation $\sqrt{x}$). This means if you just flip the ratio above and take the square root (with an adjustment for $\sqrt{x}$), you should have the predicted scaler. And you can test the prediction against actual observations.

Watch what happens. If you take the velocity based on transferred momentum only and multiply that by the predicted scaler, you get the following relationship:

$$\text{Velocity} = \underbrace{\frac{m_e c}{m_o + m_e}}_{\substack{\text{Velocity} \\ \text{(predicted by} \\ \text{transferred} \\ \text{momentum)} \\ \text{only)}}} \underbrace{\sqrt{\frac{m_o + m_e}{x\,(m_e c)}}}_{\substack{\text{Scaler} \\ \text{(predicted by} \\ \text{scale metrics)}}}.$$

When we apply this scaler, we get the exact velocity of the object every time. It never fails!

It turns out that you can determine the velocity of an object through a simple ratio of the absorbed momentum divided by the total energy content (which is what the concept of momentum would suggest) multiplied by a scaling factor (which is what scale metrics predicts).

This is a nice outcome and further evidence that scale metrics may be on the right track. Let's see if we can keep this momentum going as we move into a discussion on the equivalence principle.

18

THE EQUIVALENCE PRINCIPLE

It has been said that there are as many versions of the equivalence principle as there are authors that have written about it. For just a moment, I was hesitant to introduce yet another interpretation. I asked myself, "Is it really needed?" And I came to the conclusion that, yes, we will need a scale metrics equivalence principle. To get to the heart of this, we should start with a historical perspective.

The equivalence principle is based on the observation that all objects fall at the same rate within a gravitational influence when released from the same height. It does not depend on mass, physical composition, or any other properties of the object. They all fall at the same rate. This has been known since the 1600s, during the time of Galileo. What was not known was why.

In its earliest and most basic form, the equivalence principle established an equality between gravitational mass (the property of mass that causes gravitation) and inertial mass (the property of mass that influences its motion when subjected to a force). These two ideas of mass are not automatically the same. So over the centuries, there has been much interest in trying to determine whether they are indeed equivalent. Any tiny difference would be enough to show that the equivalence principle is flawed and that gravitational and inertial mass are not absolutely equivalent. In the early

FROM FALLING APPLES TO THE UNIVERSE

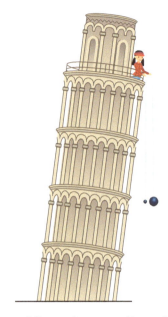

Both objects fall together regardless of their size, mass, or composition.

1900s, Loránd Eötvös conducted the most rigorous experiments at the time using a torsion balance and found to a great degree of certainty that both inertial and gravitational mass are equivalent. For most, the subject was settled, yet there are still those who investigate the equivalence principle. Today most of that work lies at the quantum level to see if any violations in the equivalence principle can be found within the tiniest scales of nature.

Yet even if you accept that this subject is settled, there is still the question of why. After all, it seems strange that a bowling ball and Ping-Pong ball should fall to the ground at the same speed. But that is actually what will happen (again, if you ignore air resistance). They will fall at the exact same rate and hit the ground at the same time if released from the same height. An experiment similar to this was actually done in 1971 on the moon by Apollo

THE EQUIVALENCE PRINCIPLE

15 astronaut Dave Scott. The lunar atmosphere is much thinner and air resistance can be ignored. Here's what happened: a feather and hammer dropped from the same height hit the lunar surface at the same time. Kind of crazy stuff, but well documented.

Einstein turned the table on this by introducing an idea that would have been strange if gravitational and inertial mass were not equivalent. This was the Einstein equivalence principle (EEP) introduced in 1907. Einstein stated that one cannot tell the difference (within a small region of space-time) between gravity and acceleration, essentially asserting that gravity and acceleration are the same. This was quite a bold statement and was viewed as a wonderful observation and part of the brilliance of Einstein. In Einstein's mind, this was critical to the development of his theory of GR.

Here is what Einstein suggested: If I am on the surface of Earth, I feel my weight. However, if I were far away from Earth (far away from gravity) in a spaceship that was accelerating, I would also feel a force acting on me like gravity. Now if I choose a rate of acceleration that provides the same strength as gravity, then I cannot tell the difference between being pressed against the surface of Earth by gravity or being pressed against the floor of my accelerating spaceship.

Imagine you're out for a spacewalk and standing on the outside of your spaceship. The outer surface of your spaceship looks just like Earth except that it has little to no mass. If this massless version of Earth is accelerating in your direction and you are standing on its outer edge, your weight will feel exactly the same as it does on the real Earth. The two sensations are indeed equivalent!

In scale metrics, acceleration is always felt and prevents you from following your natural and preferred path of motion as determined by the value of the metric. The

FROM FALLING APPLES TO THE UNIVERSE

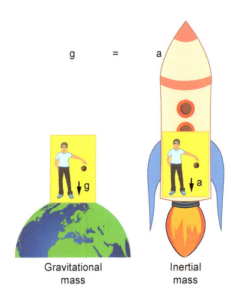

equivalence between acceleration and gravity is obvious and simply a matter of semantics. There is no mystery as to why they are the same. If the metric is not changing, an object will stay motionless. If the metric is changing, an object's natural and preferred path is to fall in the direction of the gravitating mass. But this falling movement has absolutely nothing to do with a force. Remember, the person in free fall does not feel any force.

It is only when this free-fall motion is physically blocked by the surface of Earth that an object is prevented from following its natural and preferred path. By definition, this is acceleration. Just as I suggested in chapter 2, acceleration does not have to be associated with a change in velocity. Rather, acceleration can be viewed as that which prevents an object (forces it) from moving in the way it wishes to move (or not move). When you feel pressed against the surface of Earth, you are indeed accelerating. So the equivalence between gravity and acceleration is clear. They are merely two different words describing the exact same phenomenon.

THE EQUIVALENCE PRINCIPLE

However, the same cannot be said for an object that appears to be falling toward you. When you are accelerating in a spaceship, you—and the object—are on the same map (your metric does not change).

When you are pressed against Earth, you and the object falling toward you are both experiencing a metric with a gradient that changes in the radial direction. These two situations, unlike the exact equality between gravity and acceleration, are not equivalent. Why? Because the object that is stationary in a constant metric never experiences an increase in energy. It is motionless in a uniform metric. It seems to be accelerating toward your spaceship because you interpret your acceleration to be stationary, and you interpret that a force is being applied to you that is equivalent to that of Earth's gravity. Since you experience a constant value for gravity, g, the changing velocity of the object that appears to be falling toward you is also based on that same constant value of g. This means it appears to be gaining velocity, yet its mass remains unchanged because it is motionless in a uniform metric. That would mean this object can achieve velocity without gaining any energy content, but that is physically impossible.

Now some will argue that the observer in the accelerating spaceship will indeed perceive the "falling" object to have gained energy. But this argument does not work either. If the observer perceives the "falling" object to gain energy, then its value of g must be decreasing, as it becomes more difficult to move an object with an increasing energy content (that of both the rest mass and the mass associated with the increased kinetic energy). But this does not happen; the observer from the spaceship must maintain a constant value of g to establish the equivalence between acceleration and gravity. Therefore, he must also perceive the object falling with a constant acceleration of g.

This is generally overlooked because the change in energy content over a tiny region of space-time (an infinitesimal interval) is so extremely small that it is assumed to be OK to simply ignore it. It is a lot like saying you can view your local surroundings to be flat even though you are on the surface of Earth, which is a sphere. But let's say you just bought a new car, and glued to its window is a sheet that states the weight of your vehicle. So if I were to ask you the weight of your car, you would be able to tell me. But wouldn't you have to account for the weight of any gasoline in the tank? Probably not; the tiny weight from a small amount of gas has little impact on the weight of your entire car. So you can ignore it, right? But you can't drive your car off the dealership lot if the gas tank is completely dry. You need energy to create motion, and it is physically impossible that a car with no gas can still be driven. This is entirely different from claiming that the weight of a tiny amount of gas can be viewed as insignificant in relation to the entire weight of the vehicle. It is never acceptable to claim that an object can move (even a tiny distance) without accounting for the energy that allows it to move.

To address this, scientists have moved to a concept called general covariance as a subtle transition from the EEP. According to GR and the concept of general covariance, objects are following a geodesic in curved space-time. And since all the objects are following the same geodesic path, it makes no difference what the mass, physical properties, or composition of the falling objects are. They will all appear to be falling at the same rate. This we all agree on. Scale metrics makes the same case: all the objects are moving through the same changing metric and will therefore all fall the same. But GR and scale metrics make different predictions on the value of their respective metrics (chapter 14). The metric from the curvature of space-time in GR has baked into the

THE EQUIVALENCE PRINCIPLE

mathematics the notion that acceleration and gravitation (not gravity) are equivalent. This requires that it be physically possible for an object to gain a tiny amount of velocity without increasing its energy content. Scale metrics suggests something different based on the realization that a change in energy content is always required for any change in velocity. No exceptions!

I have made my point, but I have still not clearly articulated what the scale metrics equivalence principle is. Let's take that up in the next chapter.

19

SCALE METRICS EQUIVALENCE PRINCIPLE

To summarize, there are several different ideas that all contribute to the concept of the equivalence principle:

- "Gravitational and inertial mass are equivalent." True, based on all experimental observations to date.
- "All objects dropped together from the same height fall at the same rate and hit the ground at the same instant." True, ignoring air resistance.
- "You cannot tell the difference between acceleration and gravity." True, generally limited to a small region of space-time.
- "You cannot tell the difference between the path of an object falling to the surface of Earth via gravitation and the path an object appears to fall due to your acceleration in the absence of gravitation." Not entirely true. An object's energy content influences its velocity (even across an infinitesimal interval).

The scale metrics version of the equivalence principle addresses this last bullet point by examining the kinetic energy required for any motion. When this is done, it becomes clear that there is a fundamental difference between gravity and gravitation. Let me explain.

If I am in free fall, I gain velocity as my energy content increases, resulting in an increased mass. The greater the energy content, the harder it becomes to make me move faster. The path that I follow is the result of gravitation. It is gravitation that determines the natural and preferred path that an object will follow in the absence of any force. This path is defined by a changing metric in the radial direction.

Gravity, on the other hand, is truly a force. It is what I feel on Earth (or any other gravitating body) when my body is pressed against its surface and indeed accelerating, because my body is prevented from following its natural and preferred path (since my body is being held in a position that is not the way it wants to be moving). It is the same force that I feel if I am being accelerated in a spaceship in the absence of gravitation. Both GR and scale metrics establish a true equivalence between gravity and acceleration. But gravity and gravitation are not equivalent. Gravity is a force, and gravitation simply defines the natural and preferred path of motion for a test particle in the absence of any force. Its motion is determined solely by the amount of energy present. Therefore, to truly understand the equivalence principle, we must establish a more rigorous quantitative relationship between gravity and gravitation.

To do this, we are going to start with special relativity (SR). All motion is relative—almost everyone has heard this statement. It means that for a system with two moving objects where the velocity of both objects remains constant, it is not possible to tell which is moving and which is stationary—or, for that matter, if they are both moving.

Put a simpler way, both observers can make the claim that he or she is at rest (not moving) and that the other is in motion. This is similar to the argument that both the person in the accelerating spaceship and the person on the surface of Earth can claim to be at rest within a small region of space-time. The difference is that in SR, neither party feels any force, and in GR, both feel an equal force due to gravity,

SCALE METRICS EQUIVALENCE PRINCIPLE

or acceleration. (Remember, gravity and acceleration are two words describing the same thing.)

Or, to be more precise, gravity and acceleration are the same thing applied within two different environments. When you refer to gravity, you are describing an acceleration acting on you within a gradient defined by a changing metric. When you refer to acceleration, you are describing the felt sensation of gravity when your velocity is increasing due to a force applied to you within surroundings defined by a constant metric.

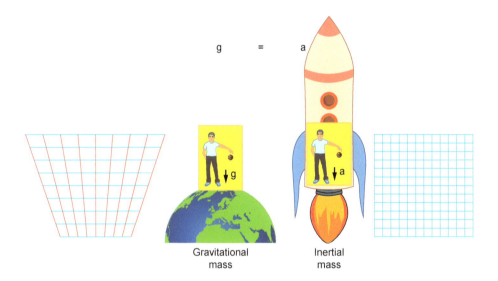

While both observers feel the same acceleration, they are making their observations from different metric backgrounds. This calls into question whether their observations are identical. More succinctly, it raises the question of whether gravity and gravitation are equivalent.

Much of my criticism of the EEP is the way in which it treats what is occurring within an infinitesimal. The EEP focuses on establishing an equivalence between two different points of view (on the surface of Earth versus in an accelerating spaceship) for a particle that is changing in velocity with time. By definition, this means that time must pass and that the location of the particle must change. EEP states that if this interval is so extremely small (an infinitesimal), we can treat the acceleration within the interval as being constant. Scale metrics suggests there is nothing wrong with developing mathematics around this idea; it is just that the mathematics will be describing something that can never occur. (Again, try driving your car with no gas in the tank. If you can accomplish this, I will concede this point.) Put another way, I am not being critical of the mathematicians; it is their job to develop mathematics. It is the physicists who have dropped the ball. It is their job to ensure that the mathematics being used is a reflection of what can actually occur within nature.

Within scale metrics, I want to change this perspective and look at the amount of energy (as opposed to acceleration) that is required to achieve the observed outcomes. Since motion is a static property, the advantage of this approach is that we can completely eliminate the infinitesimal interval altogether. Within scale metrics, we can isolate the instantaneous motion of an object whether it is moving at a constant or changing velocity. Remember, motion is defined by the value of the orientation, x. Knowing this, what I want to do is compare the motion due to gravitation (a changing metric) of a test particle with the kinetic energy required to achieve that same motion in the absence of gravitation (constant metric).

Sounds simple enough, but it gets tricky. The motion can be viewed in two different ways. The instantaneous motion is relative. Does the test particle have the property of motion, or does the surrounding space contain motion? This brings up the idea of a frame of reference. One way to think about this is to consider a tiny object, let's say a kernel of corn, on a sheet of paper. There are two ways to look at any movement. Is the kernel moving across the paper? Or is the kernel stationary and the entire sheet of paper moving past the corn? This gives you an idea of what is meant by the concept of a frame.

But now we need to be careful because the idea of space comes into play. When using scale metrics, space has mass. So now energy is required to move a frame of space. And the energy required to move this frame versus the energy required to move a single kernel of corn is significantly different. Yet if motion is relative, there must be some type of relationship between the energy needed to move the frame and the energy needed to move the kernel of corn. This becomes central to the notion of the scale metrics equivalence principle.

To start this exploration, we first need to understand what is meant by moving the frame of space. It is probably unrealistic and certainly unimaginable that the entire universe is being moved past a tiny stationary particle. However, do I really need to move the entire space of the universe? The answer is, not really. Science is limited to what we can actually observe. I have no way of knowing how things are moving on the other side of the universe. I can only know what I can see within the duration of a particular interaction. Since what I can see is communicated at the speed of light, only so much of the universe can be directly observed during any particular interaction.

For example, let's say you are in a room with no windows. You walk across the room at a constant rate and see that you have moved from your desk to the copy machine. Yet realizing that all motion is relative, you ask yourself if it is possible that you remained stationary and that the entire room moved past you. This is what SR suggests is possible. Now I ask the question: Does it matter what the space outside your room is doing? Did your entire town need to move? Your entire state, the whole country, or the entire universe? Absolutely not. The only observation you could make was limited to the frame of the room, the frame that you could actually observe. What was happening outside the room is not part of the observation. It is not part of science. This means that the portion of space that can be observed as a frame is limited by the distance between the interacting particles. Specifically, this turns out to be a value of $2\pi hr$. To the advanced reader, this may seem like an unusual value. After all, the area of a circle is πr^2. However, you must realize that the universe is made up of segments, not points. Refer to section 3 for a full derivation.

So now we have two possibilities. A test particle can move from one position to another within a stationary frame. Or a portion of space—defined as the frame that can be directly observed—can be viewed as moving past a stationary test particle. These two observations are relative yet have vastly different energy requirements.

There is yet a third option. What if a gravitating mass is present? In this case, you would follow a natural and preferred path toward the gravitating mass. You would feel no force acting on you while the gravitating body defined your instantaneous motion at each distance r from itself. This would be the result of a changing metric defined by a gradient as you move radially toward the center of the gravitating mass.

Your velocity would be increasing as you moved toward the gravitating mass, but at any distance r, you would have an instantaneous motion. That is, we are not concerned with what you motion was the moment before or what it will be the moment after but only what it is at that instant when you are at distance r. And, at that instant, you motion is relative. You have no way of knowing if you are the one experiencing motion or if you are at rest and the frame is in motion.

SCALE METRICS EQUIVALENCE PRINCIPLE

So now we have the tools in place to establish the scale metrics equivalence principle. In fact, a number of things from previous chapters are going to nicely come together.

First, let's make the space frame move past a stationary test particle. Let's say that at a distance of r, the space frame is moving at $0.05c$ (5 percent the speed of light). We know that the mass of this frame is $2\pi hr$. We can determine how much kinetic energy is needed to make the frame move at 5 percent the speed of light. This can be determined from the value of γ, which at 5 percent the speed of light is equal to 1.00125235 (see chapter 15). Therefore the mass associated with kinetic energy is 0.00125235 times $2\pi hr$. We also know that we can determine the velocity of the object using both momentum and a scaling factor. For this, we need to determine the value of x, which will be 0.5003134 (see chapters 15–17).

So now we can determine the velocity of the object:

$$\frac{\left(0.00125235\right)\left(2\pi hr\right)}{(1.00125235)(2\pi hr)}\sqrt{\frac{(1.00125235)(2\pi hr)}{(0.5003134)(0.00125235)(2\pi hr)}} = 0.05000000\,c.$$

A perfect match!

Now watch what happens. Let's take the mass associated with kinetic energy $(0.00125235 \times 2\pi hr)$ and place this as a concentrated (gravitating) mass in the center of the space frame. The test particle is therefore a distance equal to r away from the center of the gravitating mass. In this case, what will be the instantaneous motion of the test particle due to gravitation?

For this, I need to calculate the value of the metric b (see chapter 14).

It will have a value of 0.998749216. From here, we can use the Pythagorean theorem to determine the instantaneous velocity of a test particle in free fall toward the gravitating mass (see chapter 4). And that is $0.0500000c$.

Or I could have taken $(1 - b)$ multiplied by the scaling factor and arrived at the same value of 5 percent the speed of light.

So what does all of this say? It says there is a perfect equivalence between the *mass* associated with the kinetic energy needed to move the space frame (in which your spaceship can be embedded) past a stationary test particle and the *mass* of a gravitating body (on which you are standing) inducing an identical velocity on a test particle in free fall at a distance of r.

Why is this an improvement over the EEP? Because the scale metrics equivalence principle takes into account the energy associated with motion without ever needing to consider the tiny interval of the infinitesimal. As a result, the scale metrics equivalence principle identifies a difference between gravity (a force) and gravitation (the natural and preferred path of an object as determined by the energy that is present). And while force and energy are certainly related, they are most definitely not equivalent.

In other words, scale metrics suggests that the observer within the spaceship cannot maintain a constant force while accounting for the appropriate energy requirements to observe an object falling in the same way as the observer on the gravitating mass. Why? Because the values of x and y used to determine the appropriate energy are constantly changing within each instant, and this precludes the physical possibility of maintaining a constant value of g. And without a constant value of g, the observer in the spaceship will not experience the same constant force (felt as weight) as experienced by the observer on the gravitating mass.

To put this in a real-life experience, I once gave an entire presentation on the concept of Riemannian geometry in the woodshop. I will spare you the details, but the essence of the presentation was that most carpentry instructions are based on the idea that your lumber is perfect (without any warping, such as bowing, twisting, or crooks). The reality is that if you buy a long-enough board, it will indeed have some warping. When you get back to the woodshop, you had better make some adjustments to the instructions. If you follow them exactly as written and apply them to a warped board, you will end up with something that is not exactly what you had hoped to build.

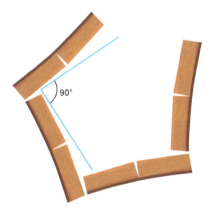

If I intended to build a square picture frame by connecting the joints at 90 degrees, I would not get the desired outcome if the boards were warped.

When applying this story to mathematics, I am not saying that mathematicians cannot develop a relationship that integrates a constant (perfectly flat) value of g within each infinitesimal. What I am saying is that the mathematics will be describing a situation that is physically impossible to achieve with a gravitating body of constant mass. Therefore—just as in the woodshop—the outcome will be different from what we had expected. Refer to section 3 for more details.

You might be thinking that the test particle in the previous example was moving quite fast. Five percent the speed of light is much faster than any observations in everyday life. Does the above approach work for everyday situations?

It does. It once again shows that the classical equations of motion are not valid (except when no motion occurs). Even with the tiniest of velocities, if you determine the values of x and y, you can determine the velocity of the space frame moving past a particle. And since motion is relative, this is the same velocity as a test particle moving toward a massless object at rest for each distance r. You now see why all objects fall at the same rate due to gravitation. We never needed to account for the properties of the test particle to determine its relative velocity to the space frame. And this motion is the same whether it is achieved from kinetic energy (which has a mass equivalency) or from a gravitating body consisting of an equivalent amount of mass.

So where does this leave us? Scale metrics suggests the following:

- Gravitational and inertial mass are equivalent.
- All objects dropped together from the same height fall at the same rate and hit the ground at the same instant.
- You cannot tell the difference between acceleration and gravity.
- Identical motion is induced by either kinetic energy acting on a space frame of radius r or by gravitation (from the energy content of a gravitating body whose mass equivalence is that of the kinetic energy) acting on a test particle at a distance r. This motion is determined by the natural and preferred path of the object and is never due to the force of gravity—demonstrating that gravity and gravitation are not equivalent, as required by the EEP.

20

FURTHER DEVELOPMENT OF GENERAL COVARIANCE

In chapter 18, I mentioned that the idea of general covariance has been used in place of the EEP. General covariance is a powerful conceptual idea that incorporates EEP. That is, if you use general covariance, then you automatically satisfy the conditions of the EEP. So what is general covariance?

It is a theoretical concept that humans are the ones that create coordinate systems. I can use Cartesian coordinates (x, y, and z), or I can use polar coordinates (angles and distances) as two examples. There really is no limit to the variety of coordinates that might be developed by humankind. Furthermore, I can define any coordinate system to be at rest, moving at constant velocity, or moving with a changing velocity (within a small region of space-time), and none of that should have any impact on how I view the laws of nature. The idea of general covariance is that nature does not care what coordinate systems are developed by humans. Nature is nature, and therefore the form of the expression describing nature should be the same regardless of the coordinate system we started with. This introduces the mathematical concept of tensors, which are used to develop a common form to our expressions for

FROM FALLING APPLES TO THE UNIVERSE

the laws of nature. This is a very attractive and seductive argument.

General covariance is based on the notion that a coordinate system is a complete abstraction. There are no coordinates to the universe. We just pick a system to use to help us understand and measure the universe. But none of the systems are better than any other, and none of them are absolute—in fact, according to general covariance, none of them are even real. They are all fabrications, abstractions of the human mind.

I bring this up because those more advanced readers will simply state that much of my focus has been on the EEP, which may be interesting but does not in any way challenge the notion of general covariance. There are no coordinates; we just make them up, and therefore whatever we start with should not have any impact on the form of the expression we develop to describe nature.

How do you argue with this?

Let's go all the way back to chapter 5, where we defined the concept of the metric. A simple, perhaps rather innocent statement was made at the time: "The distance between two free particles defines the metric. In this case the metric is physical, unlike in GR, where the metric is an abstraction. That is, the concept of the distance between two adjacent points of space has no real physical meaning in GR, yet in scale metrics it is real and physically present."

In scale metrics we do indeed have an absolute coordinate system. It is called space. One of the biggest gaps in scientific knowledge may well come from our willingness to dismiss the incredibly important role of space. Remember chapter 5, where I stated: "If you take nothing else away from this book, at least walk away with the notion that our current perceptions of space may be completely wrong. Space is fundamental to our understanding of gravity, gravitation, and

FURTHER DEVELOPMENT OF GENERAL COVARIANCE

the nature of the universe. It is the entity from which distance, mass, and time are derived, and it is critical to our understanding of matter, motion, and the formation of the universe. Put bluntly, space is more complex and robust than we have ever given it credit for."

To help you understand this, here is an interesting and somewhat mind-blowing exercise. I want you to define your own personal version of the universe using scale metrics as a guide. Select a square surface (a piece of paper, countertop, coffee table, windowpane, picture frame, etc.) that will define your notion of the universe. We will find out that we can each pick a unique surface of any size, and it will have absolutely no impact on how each of us ultimately views the universe.

It makes no difference if you choose a small surface and I choose a large one, because the size of the universe is not defined by how big your initial square is. Rather, the universe is defined only by the number of free energimes within your square. So regardless of what you think the size of your square is, it represents absolutely no space unless free energimes have been emitted.

The square that you just selected defines the scale of the energime. This should not be surprising, because it is the very same concept that we applied in chapter 17 to determine the value of the scaling factor that is multiplied by momentum to determine velocity.

What is the advantage of the scale metrics universe? For one, it prevents you from thinking about space as existing, or expanding, into other already preexisting space. Once you wrap your head around the concept that there is no space without the presence of free energimes, everything else falls into place.

We are all allowed to make the initial square of our universe anything we want, and that square has nothing to do

FROM FALLING APPLES TO THE UNIVERSE

with the size of the universe. This might not make any sense to you. But if you are struggling with this notion, it is because you're trying to think about space in the presence of already preexisting space. The universe does not work this way using the scale metrics model. This is a very hard mindset to break, but it is fundamentally necessary if we wish to make any sense out of the concept of space. The only space that exists is that which is created by the phase change between bound and free energimes within your square. Everyone can make the square as big or small as desired, and each square defines the unity value of the energime scale (the map on which the energime operates).

This raises an interesting question regarding whether free energimes actually move. The energime can be diagrammed using a vector in static time. But how would this translate into movement over an interval of time? The energime may indeed not move because—in fact—it has nowhere to move. Its motion vector is defined by the dimensions of your square, and there is nothing outside your square for the energime to move into. So does the energime move? That may be a question in search of an impossible answer. What do I mean by that? If all energimes are identical, then it is impossible to tell if energimes are swapping positions with each other or simply remaining stationary.

Think of it this way: if you have identical marbles at each position within a grid and these marbles exchange positions, can you tell? What if you had a friend mix them up (or not) while you were out of the room. Could you tell upon your return if any marbles had been moved?

Therefore it makes no sense to talk about the movement of an energime, because we can't assign any position on the grid to any specific energime. They are all interchangeable. The concept of any specific energime's movement across the grid is meaningless. The property we can understand is the motion vector for the energime, as we realize that the combination of all free energimes results in a universe at rest. (Just as adding all the motion vectors for an object at rest cancels them out.) Therefore we must restrict our conversation to the energime's motion vector in static time.

We have no way of knowing whether individual energimes are swapping positions or are stationary within their particular grid location. This plants an exciting seed for the topic of wave-particle duality. If—at the most fundamental level—I cannot tell the difference between moving energimes and stationary energimes, is there really any difference between wave motion and particle motion? Perhaps wave motion and particle motion are indeed equivalent, an integrated wave-particle duality. This stands in contrast to the Heisenberg uncertainty principle, which drives a permanent wedge between particle and wave motion. Observers determine, depending on how they set up their observation, whether they will observe the properties of a wave or a particle. Yet scale metrics suggests that all motion may be equally thought of as both wave and particle at all times. Something to discuss on another day, perhaps.

Regarding general covariance, within scale metrics I do indeed have a framework that defines an at-rest coordinate system. It is defined by the free energimes present in the universe, with each energime marking a position on the grid of the large-scale structure of the universe. Within this coordinate system, I can tell the difference between an object at rest, an object moving at constant velocity within the grid, or an object moving at a changing velocity relative to the grid. Within scale metrics, the concept of general covariance does not hold true.

Nature has provided us with space as an absolute coordinate system!

21

BLACK HOLES: ARE THEY REAL?

I don't know if you were patiently waiting to get here, but we are finally prepared to ask the question: Are black holes real? Of course this all depends on how you define a black hole. Generally speaking, the single most identifiable property of a black hole is the event horizon. Once you cross this boundary, you can never escape the intense gravitation. You are forever trapped. This is the property that science fiction movies have the most fun with (and take the most liberties with).

What is interesting is that as you fall toward the black hole, you have no knowledge that you crossed the event horizon. It is only as you try to leave and move back outward into space that you realize you are forever trapped. Equally interesting, and contrary to common belief, at distances beyond its event horizon there is nothing special about a black hole. It is simply a gravitating mass like any other. Yes, it is massive, but it has no special gravitational properties. It does not suck everything into it any more than any other gravitating body does; it alters the surrounding metric in the same way, causing objects to move toward its surface.

We have the equation from GR that I shared earlier (chapter 14):

$$\frac{ds}{dr} = \frac{1}{\sqrt{1 - \dfrac{2GM}{r}}}.$$

As you can see, if the value of $2GM/r$ is equal to one, then the denominator goes to zero and the equation blows up. This means if $r = 2GM$, the equation will blow up. So this distance of $r = 2GM$ defines the event horizon and becomes the special threshold that, once you cross over it, can never be crossed again.

Interestingly, this event horizon radius is often referred to as the Einstein radius or the Schwarzschild radius, named after physicist Karl Schwarzschild, but neither of these men ever suggested the existence of an event horizon! Einstein spent many years attempting to develop an explanation as to why an event horizon could never actually form. Schwarzschild died shortly after solving the Einstein field equations, but his solution held no possibility of an event horizon. It was the famous mathematician David Hilbert who derived the version of the equation that is used today and that still bears Schwarzschild's name. It is amazing how few physicists ever reference this or even appear to be aware of it.

The event horizon is a direct result of GR, and for an object falling straight into the black hole, it would be governed by the change in the metric as defined by the above equation.

Using scale metrics, it is defined differently, by the equation

$$b = \frac{1}{1 + \dfrac{GM}{r}}.$$

Remember chapter 14, where I demonstrated that you cannot tell the difference between the GR equation and the scale metrics equation for weak gravitational influences. But at strong gravitational influences, such as those for black holes, they would differ. It turns out that it is not so hard to bring GR into agreement with scale metrics. All one has to do is replace the two in the GR equation with value of $1/xy$ that was developed to provide an exact equation of kinetic energy.

$$b = \frac{1}{1 + \dfrac{GM}{r}} = \sqrt{1 - \frac{GM}{x\gamma r}}.$$

I promised that we would establish a true equality between GR and scale metrics in chapter 14.

The above equation works at any gravitational strength. But how do I know that GR should be adjusted? Why not provide an adjustment to scale metrics to bring it into agreement with GR?

It turns out that the two in the GR equation can be traced back to the 1/2 in the classical kinetic energy equation:

$$\frac{ds}{dr} = \frac{1}{\sqrt{1 - \dfrac{2GM}{r}}} \qquad\qquad E = 1/2\, mv^2$$

However, I have already shown that the xy version is a better equation suggesting:

$$\frac{ds}{dr} = \frac{1}{\sqrt{1 - \dfrac{GM}{x\gamma r}}} \qquad\qquad E = x\gamma\, mv^2$$

We can go even one step further. We can discuss ad nauseam whether EEP or general covariance are valid. Or we can ensure that the local space of each infinitesimal is perfectly flat. (Remember, this is required to utilize Riemannian geometry.)

All we need to do is provide a slight tweak of additional gravitational mass within each infinitesimal to ensure a constant value of g for a changing velocity. Why is this necessary? Because as the object gains velocity, it gains kinetic energy, and that makes it harder for it to continue gaining velocity at the same rate. If we wish to ensure a constant rate of change for the velocity (even within the tiny distance of the infinitesimal), we *must* continuously add mass to the gravitating body. There is no way around this if we wish to ensure a constant rate of change for velocity. And when we do that, we bring GR into complete agreement with scale metrics. Further, when this happens, it turns out that an infinite amount of mass is needed to reach the condition where $r = GM/_{xy}$.

Since there is no amount of mass that is greater than infinity, it is impossible to cross over such an event horizon. This is not a problem for scale metrics, because the event horizon occurs at the very center of the gravitating mass. It turns out nature did not get lazy. It had built into the design all along a provision that prevents a situation where an event horizon can form, just as it made sure that provisions were in place to ensure that the speed of light cannot be exceeded. That is, you need an infinite amount of energy to reach the speed of light, and you need an infinite amount of mass to generate an event horizon. See section 3 for more details.

It turns out nature had it covered all along!

22

"I DON'T BELIEVE YOU: WE HAVE SEEN BLACK HOLES"

Some readers are going to push back on the evidence provided in the previous chapter. After all, black holes have been confirmed, right? We know that the center of the Milky Way galaxy has a black hole named Sagittarius A*. There is other strong evidence of black holes; the idea is that every large galaxy has at its center a massive black hole. We have even seen an image of a black hole as modeled by the Event Horizon Telescope. How can I say they are not real?

I hear you loud and clear, but let's think about this. Do black holes exist after all? Again, it depends on how you define a black hole. If you are defining it by a specific event horizon, then I stand by my statement—black holes do not exist. But if you are defining it as light that cannot reach us, then that may be a different story.

It is certainly appropriate to consider that the center of many galaxies may have a massive object. But that is all these objects are; they are massive gravitating bodies that rule their galaxies in a way similar to how our sun rules the solar system. Their gravitation is fully explained by the model of scale metrics. And, yes, it is very possible that we cannot see the light emitted by these massive gravitating bodies. So what is going on?

FROM FALLING APPLES TO THE UNIVERSE

Humor me with two more stories. This story is called "Lost in the Woods."

I am lost in the woods, and I am calling out for help. Whether you can hear me depends largely on how far away you are. The sound dissipates as it travels farther away, and you may simply be out of range to hear me yelling. That is one possibility.

Now let's look at another version of the story. I am walking in the woods, and I see a little shed. I am curious, so I walk inside and the door closes behind me. Nothing looked out of the ordinary as I walked into the shed, but once inside I realize that the shed is totally soundproof and that the door is locked. I yell for help, but you cannot hear me, because no sound can pass outside the shed. You could be walking right by the shed, and you would still not hear my pleas for help. If only I had stayed just one step outside the shed, you could hear me. But one step into the shed the door closes, and I am forever trapped, unable to send my pleas for help.

So which is it? You would need to conduct an experiment to answer that. Let's say you have a general idea of my location. So even though you cannot hear me, you walk in the direction that you believe is toward me. If I am screaming outside the shed, eventually you will hear me. It will be very faint at first, and then stronger and stronger as you move toward me. But if I am inside the shed, walking toward me provides you with no benefit. You can keep moving closer to me, but you would still never hear me.

What is the point? To truly verify the existence of a black hole with a specific event horizon, I would need many data points, each one closer than the last to the perceived black hole. But I am stuck on Earth, or at the very best with measuring devises located within our solar system. I have no way of knowing from my current (and relativity fixed) position what I might see if I were able to move closer to

"I DON'T BELIEVE YOU: WE HAVE SEEN BLACK HOLES"

the perceived black hole. I cannot conclusively prove that a black hole exists (as defined by a specific event horizon) from a single position within space. But I can, as is often done, apply the mathematics of GR to state that the event horizon is real. But GR is a model; it is not reality. Using GR as proof of a physical phenomenon without sufficient observational evidence is no proof at all.

Just think about this. Granted, science is not always intuitive, nor does it always follow common sense. Nonetheless, try to picture a situation where on one side of the event horizon you can escape a great distance away, but just one step over the event horizon and you are trapped forever. Perhaps you can come up with examples that might support this. Here's one: if I am standing on the edge of the Grand Canyon, I can turn around and get back to my car in the parking lot and go home. But one step further and I fall off the edge, and things drastically change. I am not getting back to the parking lot anytime soon, that is a given. Yet there is a problem with this example. You see, when you cross over the event horizon, there is nothing unusual that happens to you. It is only when you turn around and try to get out that you realize you are trapped. So the example above only works if you can step off the edge of the Grand Canyon and remain unaware that anything significant just happened until you try to turn around to get back to your car.

The next story is called "Falling into the Well."

I fall into a dry well and fortunately do not break any bones. It was easy to fall in, but getting out will be a different story. That will depend on how much energy I have. One version of the story ends tragically: I claw and climb all the way back to within one meter of the top, but then my muscles give out and I fall back to the bottom of the well. If the well had only been one meter shallower, I would have made it out. So whether the story ends in tragedy or happiness

FROM FALLING APPLES TO THE UNIVERSE

depends entirely on the depth of the well (and my physical strength on that given day).

But what if when I fall into the well, some trap door above me slams shut? In this case, I am forever trapped behind the door and never have any chance of making it out. But does it make any difference to my loved ones? Either I made it out of the well or I did not. The specifics are of little comfort.

So what does scale metrics have to contribute to this discussion?

Scale metrics suggests there is no soundproof shed or slamming trap door—no such thing as a specific event horizon. But the idea that light cannot reach us is a very real possibility. Does the light have enough strength (energy) to climb out of the well? That depends on how deep the well is. But whether the light was stopped by a trap door near the bottom of the well or whether it reached all the way to within one meter of my eyes and then fell back down makes absolutely no difference as to the observed outcome. Either I saw the light or I did not. All we know conclusively is that light from some distant mass is not reaching us here on Earth. That is all we know.

Now from the perspective of Newtonian mechanics, potential energy would be converted to kinetic energy, and that same kinetic energy could be used to return you to the same position that you started from (ignoring the fact that no energy transfer is 100 percent efficient). If the light fell in, it should have enough energy to climb back out. This reasoning is based on classical theory. In scale metrics, the metric is a gradient that changes as you move toward a gravitating mass. The scale of your map keeps changing as you move inward and is much different on the surface of a gravitating mass than it is far, far away. The energy available to climb out of the well is determined by my local metric, which is a much different metric than the one that exists at the top of

"I DON'T BELIEVE YOU: WE HAVE SEEN BLACK HOLES"

the well. In scale metrics, it is very likely I will never make it out of the well. But, again, this is no proof of a rigid event horizon. It does not prove that if I stay on one side I can escape or that just by moving closer to the gravitating mass I am forever trapped.

Scale metrics suggests that at one point the light will reach Earth but also that if you just move one interval farther from the gravitating mass, its light will fall short of reaching Earth. Yet if my location where closer to the gravitating mass, I would be able to see its light.

So are event horizons real or not? That will require space travel far beyond Earth to find out. But to say that event horizons exist simply because the mathematics of GR suggests they do is probably not a very good approach. If scale metrics ends up providing a decent model of reality, then the event horizon probably does not exist.

So how good is scale metrics? If we want to determine whether scale metrics is a good model, we need to better understand the idea of an infinitesimal interval.

23

LOCAL VERSUS GLOBAL GEOMETRY

When I talk about a global geometry being different from the local geometry, it can seem a bit abstract. In addition, the idea of an infinitesimal is also somewhat hard to grasp. I want to explore this with you and remove some of the mystery behind the connection between a local geometry (on a very small scale) and a different global geometry (as seen on a large scale). In the process, we may also learn more about why scale metrics is perhaps a better approach.

When you hear about using Riemannian geometry to create a global geometry from a local geometry, it may sound like a special technique. However, in actuality you can create any global geometry from any local geometry as long as you know the rules on how to connect the pieces together. This process is really not abstract or mysterious at all. There are many real-life applications of this that start all the way back in childhood. For example, think of all the structures that can be built with building blocks. One of the more extravagant examples is LEGOLAND, a whole display constructed with objects that are basically square or rectangular. The elephants made from LEGO blocks on display at the LEGOLAND Billund resort in Denmark look almost lifelike and provide a good example of using a local geometry (blocks) to create a much different global geometry (an elephant).

Credit: Alexis Duclos

So when you think of it this way, the idea of a global geometry constructed from a local geometry is pretty easy to comprehend. Now let's push this idea. The smaller the building blocks are, the more realistic you can make your creations. That is, the smaller the blocks are, the more you can create the appearance of "rounding out" the edges so that the large-scale structure is smooth and continuous. The idea of an infinitesimal is that if I make my blocks small enough, I can actually achieve a smooth appearance. This is where things get somewhat abstract and perhaps confusing.

LOCAL VERSUS GLOBAL GEOMETRY

Let's dive deeper into the mathematical concept at play. This is all related to a concept called the limit. Using our example of building blocks from above, it would go something like this: Let's say I want to build a person's face with blocks. There are lots of details that go into a face. If I start with huge blocks, I might be able to build something that resembles a face, but it will be lacking all kinds of detail, and the edges of the blocks will be very obvious. Now if I use smaller blocks, I can achieve more detail without having the transition between blocks so obvious. I can keep going with this argument, suggesting the use of smaller and smaller building blocks. Each time I transition to a smaller set of blocks, I end up with a more detailed and realistic-looking face. Now in mathematics I can always come up with the notion of a smaller block than the one I am currently using. And therefore I can keep this process going in a never-ending loop, with each rendition resulting in a smaller block. The idea is that if I take this argument to the limit, then the blocks that I would ultimately use (at the limit) represent an infinitesimal, and I would be able to create an exact likeness of the face I was attempting to display.

The local geometry must be very, very tiny. I then need to know how to connect these tiny pieces to get the result I want. After all, even if I have tiny building blocks, I cannot reconstruct a person's face without the instructions on how to combine the blocks. As a child, your building-block set probably came with a whole manual on how to build different things. Of course with time you learned how to build things without instructions, but you still knew how the pieces could be arranged to make something that globally was different from the local geometry of the simple block.

So here is the million-dollar question: Can I turn a square block into a curved surface? The answer is no. The properties of the block remain the same; it is simply the way you connect the blocks that provides a structure that has curves to it.

This can be better demonstrated by looking at the mathematics of a simple geometry, the circle.

24

THE CIRCLE

A circle is a fairly basic shape in geometry that we are all familiar with. It is defined by a specific radius, which is the distance from any point on the circle to its center. As I consider larger circles, I immediately see that they are defined by longer radii. Next I consider only a small section along the circumference of the circle, which represents an arc. Each arc is also defined by the same radius as that which defines the entire circle. I see that as my circles get larger, the arc of the circle appears to be flatter and defined by a radius that continues to be longer. If I take this concept to the limit, eventually I will end up with a straight line segment defined by a radius that is infinitely long. This is indeed the way in which a line segment is often defined. It is an arc with an infinitely long radius.

Now I can also draw a perfect circle using a pencil and compass, but how do I express this mathematically? My circle drawn with a compass has a specific radius that defines it. But my circle (yes, a perfect circle) determined using mathematics is made of segments that have been arranged so that their radii (which are infinitely long) all cross at the proper location at the center of the circle. Two perfect circles, but constructed differently. One is made of the combination of true arcs, the other with the combination of tiny line segments. We know from above that if these segments are taken to the limit, they will represent infinitesimals. And when these infinitesimal intervals are connected using the correct rules, they will indeed represent a perfect circle.

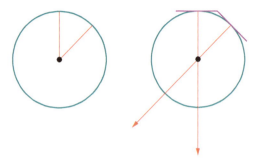

Perfect circle drawn with a compass (left).
Perfect circle constructed with infinitesimals (right).

But—and this is the important point—the rules are not interchangeable. I cannot use the rules for combining flat segments if my pieces are indeed curved arcs. Nor can I suggest that my pencil and compass are drawing flat segments at infinitesimal intervals. The two processes, while the outcomes may be the same, are not interchangeable in terms of the instructions to create a perfect circle.

So how does this all relate to gravitation? GR would have us believe that at the level of the infinitesimal, we are truly starting with flat space. This is the argument made by the EEP and also incorporated into the concept of general covariance.

Scale metrics provides a different model stating that you need to account for the changing energy content of an object as its velocity changes. Scale metrics states that the infinitesimal interval is not flat. Therefore when you apply the rules for connecting flat pieces to pieces that are indeed not flat, you will not get the desired result. In other words, you do not get an accurate description of gravitation. This is just as

I suggested in chapter 19 with my story about Riemannian geometry in the woodshop. If you use the instructions for a perfectly flat board with a board that is warped, you do not get the outcome that you wanted.

25

AN APPLE FALLING FROM A TREE BRANCH

We started this book with the question of why an apple falls from a tree. We have come far in modeling different ways of answering that question. In scale metrics, it is a changing metric that alters the scale of each local map as an object moves closer to the gravitating mass. But in all cases, we have determined the effect using a test particle that has originated infinitely far away. So this does not really answer the question at hand. When an apple falls from a branch, it is falling from a specific height (not from infinity). How then do we determine the velocity of the apple when it hits the ground if the apple did not start its journey infinitely far away?

This might seem like a simple calculation. If we were to model this off classical mechanics, we would determine how fast an object would be moving if it fell from infinity to the tree branch and then from infinity to the surface of Earth, and then we would subtract the two values to determine the interval from the tree branch to the surface of Earth.

However, this will not work, because when the apple fell from infinity to the tree branch, it arrived at the tree branch with a velocity (and therefore an increased energy content). If we stop it, we remove all its kinetic energy, and then we

FROM FALLING APPLES TO THE UNIVERSE

start fresh with zero velocity as it falls from the branch to the ground. So subtracting the two intervals will not really give us what we want.

Another approach is to determine the value of the metric at the top of the tree branch, then determine the value of the metric at the surface of Earth, and then apply the Pythagorean theorem to calculate how the value of the metric changes between the two locations. This, offhand, appears to be a more promising approach. But this is where something very interesting occurs.

It turns out that when looking up from the surface of Earth to the tree branch, we may not actually be able to determine its actual (coordinate) location. To make this easier to understand, let's take a distance that is much greater than the height of a tree branch but less than infinity.

From the surface of Earth, I measure everything I see using my local metric of mass, distance, and time. I project my local metric onto everything I measure. Let's just look at distance for now. If I am looking upward at an object a thousand meters high, I perceive that object as being a thousand meters away using my local metric. But as I go upward, within a gravitating field, the metric changes. So the location where something exists in a coordinate sense is not necessarily where I see it from a proper sense.

Therefore, I need to look at the metric as it resides in a coordinate scale to determine how fast it will be falling though a distance measured from my proper scale. Once this is done, I can accurately determine how fast an apple will be moving when it hits the ground when falling from a tree branch.

The difference between the coordinate location and the proper location is related to the value of my metric, b. Therefore, the correct equation for determine the value of b is

AN APPLE FALLING FROM A TREE BRANCH

$$b = \frac{1 + (b)\dfrac{GM}{r_i}}{1 + \dfrac{GM}{r_f}}.$$

where r_i represents the initial position of the object (the apple on tree branch) and r_f represents the final position (the surface of Earth).

However, it would be helpful if we could express this without having the value of b on both sides of the equation. With some rearrangement, we get

$$b = \frac{1}{1 + GM\left(\dfrac{1}{r_f} - \dfrac{1}{r_i}\right)}.$$

We now have an exact method of determining the change in the value of the metric for objects that are falling from less than infinity. And I can apply this metric with the Pythagorean theorem to determine the apple's velocity when it hits the ground (see chapter 4). Or I can equally determine its momentum and apply a scaler to achieve the same velocity (see chapter 17).

26

CELESTIAL MECHANICS

B efore we conclude our journey on gravity and gravitation, we need to broaden our scope just a little further. So far, we have only looked at objects falling straight down. Once we get this right, we can then look at more complex situations. In the previous chapter, we looked at the situation where an object did not fall from infinity. This provided the tools to determine exactly how an apple falls from a tree branch to the ground. It is a real-life application yet one that is rather limited in its scope.

In reality, most gravitational influences in the universe do not involve an object falling straight down. Just look at our solar system: its motion is governed largely by the mass of the sun. Yet planets are not falling radially into the center of the sun, but rather they are in elliptical orbits around the sun. Moving outward toward galaxies, we find that they, too, are moving about in a much more complex way that is governed by celestial mechanics. The point being that celestial bodies are not simply moving radially toward each other (falling directly into each other). If we truly wish to understand gravitation, we need to consider these other situations.

When you think about more complex gravitational movement, you need to consider the geometry of space itself. Within GR, a commonly used phrase is "the curvature of space and space-time." The use of the words "space" and "space-time" so closely together is sometimes confusing. At times, people get lazy and treat these as if they were interchangeable. Others think this is perhaps redundant. In fact, space and space-time are neither interchangeable nor redundant. The curvature of space and the curvature of space-time represent two totally different things.

Using GR, when we explore objects falling straight into the gravitating mass, it is the curvature of the geometry of space-time (both its spatial and time dimensions) that must be considered. But when you consider movement that is horizontal to the gravitating mass, it turns out there is also a curvature just within space itself. In other words, the spatial dimensions curve without affecting the temporal (time) dimension.

Scale metrics interprets this curvature somewhat differently. After all, the three dimensions of space are each defined by the one-dimensional entities of mass, distance, and time. In scale metrics, we cannot separate time from space. This takes a bit of thinking, but even an instantaneous snapshot of space (where no time passes) includes time. There is a difference between static mass and a changing mass, static distance and changing distance, and static time and a changing time. Just because a mass is not changing does not mean that no mass is present. Just because a distance may not be changing does not mean that a distance is not present. So it is with time as well. Just because an interval of time has not gone by does not mean that time is not present.

Within the notion of scale metrics, it is nonsensical to talk about the curvature of space independent of time because there is no space without time. So how should we approach this? Let's start by looking at distance as a constant metric. In the absence of a gravitating mass, the large-scale structure of the universe is simply a grid. The distance metric is a constant. If I wish for movement to occur within the same metric, I can simply draw a straight line.

CELESTIAL MECHANICS

Path of movement following a uniform (flat) metric

However, when a gravitating mass is present, the metric changes as you move radially toward the gravitating mass. In this case, what happens when an object's movement is horizontal to the gravitating mass? If you wish to "stay in your lane" or, in other words, stay on the same map (metric), you must actually travel along a curve.

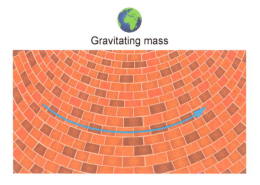

Path following a constant metric within the presence of a gravitating mass

This curvature must be considered when you calculate how an object will fall. The natural and preferred path for a photon of light moving past a gravitating body can be shown as the yellow curve below:

The yellow line shows the natural and preferred path of a photon moving from left to right as it falls toward the gravitating mass.

There are two clear factors influencing the yellow line above. First is the curvature of the blue line (representing a constant metric), and second is the inward motion of the photon (caused by the changing gradient of the metric).

Of course the example of the bricks above shows only the circular pattern of free space as emitted by the gravitating mass. Just as we do for an object falling radially inward, we once again need to consider the contribution of the gravitating mass integrated into the contribution of the large-scale structure of the universe. When this occurs, the curvature is much less extreme. It is lessened or buffered in the same manner as it is for the object falling radially inward.

For a pulse of light moving horizontally past a gravitating mass, the contribution of a changing metric that creates a radial velocity inward provides the exact same contribution as the horizontally curved path that defines a constant metric. In other words, the curvature is exactly twice what it would be if you only considered the change in the value of the metric.

It is interesting to note that this curvature in space was initially missed by Einstein. In fact, GR has no such requirement

CELESTIAL MECHANICS

that space must be curved, and it has further no stipulation that the curvature of space must be exactly matched with the curvature of space-time for a pulse of light horizontally skimming a gravitating mass. Einstein later realized this and determined that an equal curvature of space and space-time contributed to the observed bending of light as it skims past the surface of the sun. This was one of the three Einstein tests of GR, and it was verified by Arthur Eddington in 1919.

Scale metrics actually passes a much more rigorous test than GR. The only possible outcome within scale metrics is an identical contribution from both the curvature of space and the corresponding changing gradient of the metric. GR has no such requirement. It is merely a coincidence that they are the same. And, in fact, other theories of gravitation, most notably the Brans-Dicke theory, have been based on the idea that these two contributions to the curvature of light are slightly different.

GR and scale metrics are in total agreement for a weak gravitational influence, and therefore scale metrics will pass all the same tests that are proposed for GR. Scale metrics agrees with GR on the gravitational redshift of light and the perihelion shift in the orbit of Mercury. Scale metrics also agrees with GR on the EEP as long as this and all the above tests are conducted in the presence of an extremely weak gravitational influence. GR and scale metrics have different predictions for extremely strong gravitational influences because scale metrics accounts for the energy requirement to make an object move, while GR does not (at least within the infinitesimal interval). This is true for both movement that is radially inward toward a gravitating mass and more complex celestial mechanics. Once again, the fundamental reason for the difference is that GR treats gravity and gravitation as equivalent. Scale metrics has demonstrated that they differ.

The field of celestial mechanics is much more complex than this brief overview. While much of this lies beyond the scope of this book, I want to at least touch on the subject. But with this, we have concluded our journey into the scale metrics model of gravity and gravitation.

Before we move on, let's take a moment to summarize some key takeaways from section 1.

Our fundamental goal was to find a way to describe gravity (and gravitation) using only the three dimensions of our physical awareness and the Euclidean geometry that we experience in everyday life. We accomplished this through a metric whose scale changes in flat space instead of a metric whose scale changes due to a curvature in space and space-time. In developing this idea, we learned a number of things (as viewed within the framework of scale metrics):

- Space is much more complex than we may have thought and is manifested through a phase change from bound to free energime states.
- Space may be more appropriately defined by segments rather than points, with each energime defining the unity values for mass, distance, and time.
- The large-scale structure of the universe plays a significant role in understanding gravity and gravitation. We cannot separate a gravitating body from the large-scale structure of the universe. As such, the universe may be the size it is so that we can exist as we currently do here on Earth.
- Motion is a static property of matter. That is, an object does not need to move (change its location) to possess the property of motion.
- The energy of motion (kinetic energy) cannot be accurately expressed using a classical equation that does not account for the requirement that energy content must increase if an object gains velocity.

- GR, the EEP, and general covariance all require an object to change velocity (across an infinitesimal interval) without a corresponding change in energy content. This is the single and most fundamental difference in determining the value of the metric as defined by GR and scale metrics.
- Space provides an actual (physically present) coordinate system by which all events play out. Therefore the EEP and general covariance—while mathematically elegant—do not hold up as valid descriptors of nature within scale metrics.
- That is, there is a difference between beautifully constructed mathematics and what is physically possible within the universe. Applying legitimate mathematics to situations that can never exist in nature does not provide an accurate description of nature.
- Gravity and gravitation are two separate entities. Gravity is a true force, while gravitation simply defines the natural and preferred path for an object moving within a gradient defined by a changing metric.
- Because of this difference, the event horizon of a black hole may never materialize.
- And, finally, all physical constants can be expressed as dimensionless numbers as defined by nature. Also, all measurements can be reduced to nothing more than a count of energimes expressed in terms of mass, distance, or time.

In section 2, we will apply these key points in our exploration into the universe.

SECTION 2
UNDERSTANDING THE UNIVERSE

27

A CHANGING METRIC WITH TIME

We are now going to take a rather large turn in our journey. To this point, the emphasis has been on understanding gravity and gravitation and finding the correct way to describe the metric that defines gravitation. Along the way, we were introduced to a much richer concept of space, the idea of bound and free energimes, and the possibility that the fabric of space is constructed with segments rather than points. This is a good start, but it does not really complete the journey. Gravitation is the major influence governing motion within the universe. As such, a good model of gravitation should help us better understand the universe. Successful models, almost by definition, should have powerful predictive abilities as well as the capability to solve problems and answer questions. Never has this been made more evident to me than when I've shared my ideas with colleagues in the business world. Their response: "Nice theory, John. But what value are you providing? What does your model do?"

Section 2 begins to address this question, as we turn our attention to how the tools developed in section 1 can be applied to better understand the universe. For starters, stop and think about the square that you selected to define your version of the universe. As time goes by, you see that more free energimes are present within your square, resulting in a

FROM FALLING APPLES TO THE UNIVERSE

constantly changing value for the local metric. Remember, the metric is determined by the position of free energimes. For example, I have chosen to use a windowpane as my square representing the universe. Now imagine looking out a glass window and through a screened window behind it. Think of the mesh of the window screen as representing the grid of free energimes that have already been emitted within the universe. As time goes by, the mesh of the screen becomes more tightly woven as more free energimes are released. Each tiny square of the mesh gets smaller as free energimes are added to space. This all occurs within the square of the windowpane that defines my version of the universe.

This suggests that we really have two ways in which the metric changes. We have already explored gravitation, which is the result of a changing metric with location. Gravitation occurs within static time. That is, time does not need to pass to demonstrate the changing metric of gravitation. You only need to change your position to experience gravitation. Now we see that if we allow time to pass, the metric also changes without a change in position. In this case, you need only a change in time. I can stay in the same location, and as time goes by the metric will change. Since the metric changes evenly all around me with the passage of time, there is no induced motion that occurs as a result. As the observer within the metric, I am not even aware of it, but that does not make it insignificant.

Rather, the metric plays a fundamental role in how we view the history of the universe. This raises the question: How do scientists view the universe? You might think they simply look at what they observe over time and extrapolate it backward to determine what the universe looked like in the past. While that is partly true, the problem is that we do not have much direct observational data on the history of the universe.

A CHANGING METRIC WITH TIME

Think of this exercise: What would you see if I asked you to look in the mirror, and then I asked you to look again in five minutes? Chances are you would tell me that nothing had changed. Next, if I asked you to imagine what you might see five years from now, you would likely say that your appearance will change. Just because no changes are visible over the short term does not guarantee that they will not change over a much greater period of time.

Let's say the universe is fourteen billion years old and that humankind has really only been able to probe the universe (at least in a somewhat quantitative way) for about 1,400 years. That represents 0.00001 percent of the life of the universe. Let us roughly say that a human life span is one hundred years. That would mean the same percentage applied to a human life would be the equivalent of 0.00001 years, or just over five minutes. So the exercise from above was actually fairly accurate. The point is, you do not see yourself age in five minutes, but that does not mean you do not age over your entire lifetime.

Scientists see much of the structure of the universe to be very stable, but that—in itself—should not provide any assurance that fundamental properties of the universe have not changed with the lifetime of the universe. And perhaps they even changed significantly. So how can scientists explore changes to the universe when they have such a small slice of its history to work with?

To do this, they look far away. Remember, light takes time to reach us. When a scientist looks really far away into the universe, that scientist in a very real sense is seeing the past. In this way, we are not limited by the short interval in which humankind has actually existed to observe the universe. But it is important to note that the fundamental tool for achieving this relationship is the constancy of the speed of light. If the speed of light is always the same, and we know how far away we are looking, then we also know how far into the past we are seeing. But how do we know that the speed of light has not changed? Simply because it appears constant over a short interval does not guarantee that it is constant across the full history of time.

Furthermore, the speed of light has tripped us up before. Many thought the speed of light was a constant and never changed (when traveling through a vacuum). But then we learned what that really meant, which is that all observers will measure the speed of light to be constant in their local setting. We know that local settings can be deceiving. I think my local setting is a flat surface, but in reality I am on the surface of a sphere, Earth. The same could be true for my local surroundings within space. What I see locally may not apply throughout the universe.

And now we find out that the metric of the universe is not a constant but that it changes with time as more free energimes are emitted. This metric is critical in determining the speed of light. As the metric changes, so must the speed of light change with the evolution of the universe, even if all local observers throughout history have measured it as a constant in their local setting.

It appears as if the fundamental tool that scientists have been using to probe the past (the constancy of the speed of light) may not be the best approach. We tend to view the universe from the metric that we currently live in and project it back in time as if it were a constant. This does not really give us a good estimate or interpretation of what the universe looked like when the metric of its large-scale structure was much different and the distance between adjacent free energimes represented a different scale than it does today. At the very least, it introduces some nuances as to what we think we see when we look far away—into the past. This will be further addressed in a number of upcoming chapters.

A CHANGING METRIC WITH TIME

In the meantime, it is becoming clear that an understanding of time and space will be critical to our exploration into the universe. We can only fully understand the universe by looking at it through the lens of passing time. And since within scale metrics time is one of the components of space, perhaps it is time to revisit the concept of space.

28

SPACE REVISITED

In chapter 7, I suggested that space could be viewed as a combination of distance, mass, and time, with each of these one-dimensional entities represented in an orthogonal (that is, separated by 90 degrees) relationship with each other. Many readers may still be struggling with this notion. At the time I introduced this concept, I had also said we would revisit it. So let's think this through once again. Now that we know more, let's see how scale metrics treats mass, distance, and time in a combined property of space.

Resistance to this idea comes is several forms:

- As an observer, you can look in each direction with the same result. Each of the three dimensions of physical awareness appears to be a different orientation of the same thing. You don't see distance in one direction, mass in another, or time along the third dimension. So the idea of space being a combination of mass, distance, and time might seem wrong to you.
- Others may suggest that the arrow of time only appears to go in one direction. We are unable to go back in time; we can only experience the forward movement of time. So time is somehow different than distance (and perhaps mass), yet scale metrics suggests they are equivalent. (Note that this is equally a question for GR and will be addressed in further detail in section 3.)
- Finally, you might note that you need four coordinates to define an event (three spatial and one temporal), so you might

ask how scale metrics can define events using only three dimensions representing mass, distance, and time.

There may also be other variations of the above, but if I can successfully address these three concerns, it will go a long way in showing exactly how our spatial awareness can be viewed as a combination of the fundamental properties of the energime expressed as mass, distance, and time.

Let's take these in order:

"All the directions are essentially the same." From an observational perspective, this is clearly a true statement. When scale metrics states that the three dimensions of physical awareness are mass, distance, and time, what is actually being stated is that space can be sorted into these three dimensions. For example, let's say for dinner tonight you are served a vegetable medley of peas, carrots, and corn. When these ingredients are combined in a bowl, the vegetable dish looks the same along any orientation (in any direction). But that does not prevent you from pulling out all the peas (which many people will affirm they have done). Or, if you wish, you could go further and sort out all the peas, carrots, and corn kernels into separate piles. The vegetable dish is made of three different ingredients, but they all mix together in such a way that they are uniform in each direction. Scale metrics simply suggests that we can pull the properties of mass, distance, and time out of the combined medley we call space.

Let's look at this one other way, with a Rubik's Cube. When a Rubik's Cube is all mixed up, it has various colors on each side. When it is solved, each face of the cube has only one color. Solving the puzzle does not change any of the properties of the cube. Regardless of the arrangement of the colors, there remain the exact same number of squares and the same number of colors. So rearranging them is simply a way of organizing the colors of the cube. There are many orientations in which the cube can exist, and each of them is equivalent in relation to the total number of squares and colors. But, if I choose, I can align each face of the cube to have its own unique color, just as I can arrange space to have a unique dimension in each of the three properties of physical awareness.

Just because space appears to be one thing does not mean the components that make up space must be singular. For example, air looks the same in each direction, but it is actually made up of nitrogen, oxygen, carbon dioxide, and other trace gases. You may say that this is not fair, because we cannot see these different molecules. But you cannot see energimes, either. Scale metrics suggests that the mixture of "ingredients" making up space is homogenous and isotropic along each of the dimensions of physical awareness. However, this does not prevent me from separating each of the fundamental properties along its own unique axis of a Cartesian coordinate system.

So yes, space is defined by the number of free energimes, and each free energime defines the fundamental value of mass, distance, and time. Within the scale metrics model, these are indeed the ingredients that define what we call space.

"The arrow of time goes in one direction only." This is part of the mystery of time. Many scientists and philosophers have weighed in over the centuries on the meaning of time. Some have concluded that time is only an illusion and does not even exist outside the human mind. One way to present their argument is to view every event as having occurred at a particular instant that at one moment was defined as the present. This makes perfect sense, but it suggests that time is simply a collection of events that we string together into what we perceive as the flow of time. It is how we string these static events together that causes problems in how we view time.

Think of it as a series of cartoon drawings placed together into a deck. Each illustration is unique and unrelated to any of the others. Yet when you flip through the deck, it creates the illusion of motion as the cartoon's events unfold from the past to the present and perhaps even into the future. Of course, you may recognize this as an early form of animation. So if this is the case, why can I only flip through the deck in one direction? Why can't I flip through in the opposite direction and watch the motion as it "moves backward" in time? And why must I stop at the present? For example, if the present is defined as halfway through the deck, why can't I flip through the remaining cards and see into the future? When using GR, what appears to be missing is the portion of the rule book that defines how these independent events mapped on a four-dimensional space-time continuum are related to each other.

We know from observation that we are unable to slide up and down the time axis and experience the past, present, or future in any way we wish. But we don't know why the rules of nature preclude this. After all, we can slide back and forth along the distance dimension. You do that every day when you leave home and go to school or work or out to run errands and then return home to where you started. You can move back and forth along distance but not time. So perhaps time is more complex and fundamental in a physical way that we simply have not yet taken into account.

How do we figure this out? One of the challenges in comprehending time is that we tend to give the word "time" several different meanings. Not to get off subject, but I truly believe that the strong reliance on mathematics has allowed scientists to get sloppy with their use of language. While mathematics is incredibly important, I propose that language is equally important and provides much more power in understanding the physical world than many scientists appear willing to accept. After all, what do we often tell young children—and perhaps some adults—when they get frustrated expressing themselves? "Use your words!" Perhaps this is something the scientific community should take to heart.

What are the different meanings of the word "time"? (Or, for that matter, mass and distance too.) I can think of a moment in time as a specific year, such as 350 BC or 1776. This type of time identifies a specific point along the entire history (axis) of time. Next, I can think of a range in time, such as a movie that runs for two hours and ten minutes, a trip that takes three hours to finish, or, for that matter, my age. This form of time represents a duration as opposed to a specific point. I am perfectly free to imagine any time period or range in time, and I can certainly think to the past, present, or future. The year 2200 is in the future! The interval between 2000 and 2200 is 200 years. This type of time is very fluid, and I can move up and down the time axis with ease.

Lastly, I can think of time as representing the present. I exist in the present (as opposed to the past or future). It is this third version of time that creates the difficulties for us. This is what leads to the realization of the flow of time, that I move through a series of present moments in which moments that have already occurred are in the past and moments that have not yet occurred are in the future. In this version of time, we are precluded from moving up and down the time axis between past, present, and future. We are "locked" into only a forward-moving flow of time. And, further, this flow of time is set at a specific rate. That is, we cannot speed up the flow of time to reach the future any faster. It must occur at exactly the rate that is defined by the flow of time. So what is the flow of time? What defines it? What sets its rate? Before we attempt to answer these questions (at least with the scale metrics model), let's look at how

FROM FALLING APPLES TO THE UNIVERSE

we would address the same questions as they relate to distance and mass.

As for distance, I can pick a particular location. For example, my house lies in a specific place. This is thinking of distance as a particular point along the axis of distance. I can also think of distance as a range. For example, the distance between my home and work. This is very similar to what we can accomplish with the first two examples of time, where we viewed time as either being a specific point or a range of values.

I can also do the same with mass. I can pick a specific mass or a range of masses, such as the various masses in a random collection of rocks. But we do not routinely think of distance or mass as having any bearing on the past, present, or future. Or, for that matter, on the rate at which moments flow from the present to the future. Yet within scale metrics, we are completely free to define the present as a distance or a mass with the same reasoning applied to the present as a time. How? Because the universe is constantly changing from bound to free energimes. And these free energimes define the radius (distance), mass, and time (as a continuously running stopwatch) of the universe.

When you leave home in the morning and return in the evening, you may be at the same location, but the universe has ever-so-slightly changed, with the emission of more free energimes making the universe bigger, more massive, and older. If you will, the grid within your square of the universe has become more densely woven. Therefore, we can think of the present as a condition of the universe as defined by the amount of space that exists. In the past its condition was different, with fewer free energimes. In the future it will also be different, with more free energimes. The rate of phase change from bound to free energimes is what sets the flow of time to what we experience as the past, present, and future. It can be thought of as an increase in the

SPACE REVISITED

amount of time, distance, or mass within the universe. Or, when the dimensions are combined, it is simply the realization that there is more space in the universe with each passing moment than before. When using scale metrics, we fill in that portion of the rule book that defines how we get from moment to moment.

You may say that this is all fine, but that while I can still experience my home every night when I return, I cannot return to the day I was born. So the question remains: Why can't we move up and down the time axis and physically go back to the past or jump to the future?

The answer to this may lie in the properties of the energime. Remember, the energime is not a point but a segment. Think of the trees in the forest. The height of the trees helps define the physical awareness of space. When I walk through the forest, I can go from tree to tree. I can move outward and then retrace my steps from tree to tree until I'm back where I started. I also have complete freedom to determine how far I walk into the forest (how many trees I interact with). Yet when I look upward, there is only one tree extending toward the sky. In this case, I have no freedom in determining the number of trees to interact with, there is only one tree with a specific height. I can climb up and down the tree, but that activity will not change the fixed height of the tree. If we define time as this upward dimension, then the height of the tree defines the present. The entire tree contributes to the physical awareness of space, yet it nonetheless offers no degree of freedom in determining the property of time. There is only one tree, and it represents the present. In the past, the tree was smaller and represented a different present that we now perceive as the past. In the future, the tree will be bigger. Right now, we have only the present as defined by the current height of the tree. In the scale metrics model, time appears to "grow" as more free energimes are emitted.

We can only experience time (as the present) as defined by the present condition of the universe.

"You need four coordinates to define an event." This is certainly the position taken by GR. You need three spatial dimensions and a time dimension to identify an event. However, within GR, these four coordinates each lie on a unique axis with each axis orthogonally separated within a four-dimensional space-time continuum. And this works fine within the framework of GR. (However, there are problems with this model that preclude it from providing any guidance on how we get from one event to another.) GR is missing the instructions on how time flows. This will be covered in more detail in section 3. Scale metrics is a different model and treats time in a different manner than GR does. Time is not a fourth dimension but rather an integral part of what defines our three dimensions of physical awareness.

How do you define an event in scale metrics? You need to define a location within space and the specific size of the universe at the instant the event occurs. Yes, it takes four numbers to define an event (three to determine the location and a fourth to designate the size of the universe), but this is all easily defined using only distance, mass, and time. Put more succinctly, you define an event by a position in space and the condition of space at that moment. This requires nothing more than the three dimensions of space.

I hope this better helps you understand the nature of distance, mass, and time as components of space within the scale metrics model. Let us now move forward in applying scale metrics to the universe. But perhaps we should start by looking at how the universe is currently perceived within the standard model.

29

A BRIEF HISTORY
OF THE STANDARD MODEL

The standard model is often used to describe how the universe has evolved and the nature of its expansion since its beginning, the big bang. It is often held up as a pinnacle of human achievement. Although there remain several unanswered questions, it is generally believed to be a good theory of how the universe began and has evolved to its current identity.

I have been critical of the standard model for several reasons. I believe that our scientific models should generally fit with observation and provide strong predictions that can be verified through new observations. However, more times than not, the standard model has fallen short of this expectation and has required adjustments to bring it into agreement with observation. To further understand this, let's take a brief walk through the development of the standard model.

In 1915, Einstein introduced the theory of GR. Einstein, along with many others during that time, assumed that the universe was static, that it was not getting any bigger or smaller but rather had a constant and fixed size for all eternity. Within the GR model, gravity was pulling everything inward, so for the universe to be static Einstein had to introduce something that would hold gravity at bay, something

FROM FALLING APPLES TO THE UNIVERSE

that would counter gravity and push everything outward with as much influence as gravity used to pull everything inward. This entity that balanced the effect of gravity was called the cosmological constant.

In 1929, Edwin Hubble and Milton Humason noticed that the universe was not static but that all galaxies appeared to be moving away from each other and that the further away they were, the faster they seemed to be moving away from us. This suggested that the universe was expanding and allegedly led Einstein to state that his cosmological constant was "the biggest blunder" he had ever made. It was now clear that the rate at which galaxies move away from each other might be completely offset by the force of gravity pulling them together. No cosmological constant was needed to explain why the universe was in balance and not collapsing into itself.

It seemed to make sense that if the universe was expanding, then if we looked back in time the universe would be smaller at earlier ages. Hubble and others were able to estimate the rate at which galaxies were moving away and therefore how much closer they would have been in the past. The simple conclusion was that if you observed far enough into the past, the universe would eventually be very, very small, and at some point would all be in exactly the same location. The idea of everything being in exactly the same point is referred to as a singularity. Now this initial singularity is tricky, but if we think about the universe an instant after it began expanding, we could have a simple model of the universe that predicts its age based on the rate at which all the galaxies appear to be moving away from each other.

There you have it! This was the beginning of the standard model, which I will refer to as the early standard model. If this is a good model, it should mimic and simulate what we observe as well as make new predictions about things yet to

A BRIEF HISTORY OF THE STANDARD MODEL

be observed. Since the standard model suggests that the universe began expanding everywhere at once, it would mean that there are now regions of space that for the first time in fourteen billion years have had an opportunity to interact with each other.

This concept is somewhat tricky because you might conclude that since the universe started as a singularity, it was in complete contact with all regions of itself from its birth. But singularities are not quite that simple. Modern cosmologists would explain it this way: We can all view the universe as if we were in the center of the universe. Each of us has a valid claim on being in that center, yet—obviously— we cannot all be right. The answer to this is that there really is no center. The universe—that initial singularity—started expanding everywhere at once within the singularity, leaving us a universe with no identifiable point of origin. Therefore, yes, distant areas of the universe have never yet been in contact with each other.

This means that light from far-off regions of space would only now be reaching us for the first time in the history of the universe. This suggests that as we look in different directions toward far-off regions, they should look different. They have never interacted with each other, and therefore they have not smoothed out any differences that may have been present in the early universe. This is what the early standard model predicts. Yet this is not what is observed.

One of the most important observations in cosmology was the discovery of the cosmic microwave background radiation (CMBR). This was first detected in 1965 by Arno Penzias and Robert Wilson. It turns out that this radiation is exactly the same no matter what direction you look. Roughly speaking, this radiation can be used to measure the temperature of the universe. This observational evidence supported an idea proposed by Newton centuries earlier named the cosmological principle. It essentially suggests that over large-enough scales the universe should look the same in all directions. But the cosmological principle does not seem to be consistent with the predictions of the early standard model.

The discovery of the CMBR along with the observations of Hubble suggested that the universe was expanding, homogeneous (uniform throughout), and isotropic (similar in all directions). Locally there may be some differences; for instance, if I look in one direction, I may see the sun, and if I look in another direction, I might see Venus in the daylight sky. It is not these small-scale local differences that we are interested in. It is the observation that over large distances the temperature and structure of the universe are essentially the same in every direction. If you look out into the night sky, even with the most powerful of telescopes, you will see essentially the same pattern no matter where you turn (if you look at a large-enough region)!

That is actually pretty amazing, because there is no reason (at least according to the early standard model) why this should occur. It is a puzzling observation. One would expect that the early universe was somewhat lumpy and uneven. With time, it could all smooth out, but that could only occur if there were time for this smoothing process to be completed. For example, if I mix hot and cold air, eventually they reach an equilibrium and the temperature averages out. But that process requires sufficient time to smooth out the differences. When we view the distant universe, we are seeing areas of the universe that have never before been in contact with each other. Strange that all these distant areas should look exactly the same!

The only way the early standard model could account for this is if the early universe was also completely homogeneous and isotropic. While there is nothing that prevents

A BRIEF HISTORY OF THE STANDARD MODEL

this, scientists generally do not like having preset conditions to their models. It would be much more palatable if the universe could start in any random way and end up looking like it does today. To address this, in 1980 Alan Guth suggested the idea of inflation. According to this idea, the early universe went through a rapid expansion that smoothed everything out at its earliest age. The theory of inflation suggests that a very small region of space expanded unbelievably fast (many, many times faster than the speed of light). This rapid expansion would ensure that regardless of the initial state of the universe, the portion of the universe visible to us today would be homogeneous and isotropic. This would explain why the universe appears the same in all directions without having to stipulate any initial conditions to the universe at its birth.

The point I wish to make is that the early standard model did not predict a smooth universe. It did not agree with Newton's cosmological principle, which seemed to be supported by the discovery of the CMBR. The early standard model did not accurately mimic and simulate our direct observations. Put bluntly, the model was wrong. So the model was modified to bring it into agreement with observation. That is not how strong models are supposed to work. A good model does not require modifications to bring it into agreement with observation. Rather, it makes predictions that are verified through new observations!

After this inflationary epoch, which lasted only a tiny fraction of a second, the expansion returned to a rate more consistent with that predicted by Hubble. It was hoped that with this one correction, all would be good. However, more modifications were going to be needed. Next it was noticed that there was not enough mass in most galaxies to account for the gravitational motion of many celestial bodies within the galaxy. This suggested that there was a type of mass we

FROM FALLING APPLES TO THE UNIVERSE

were not able to observe. This ushered in the idea of dark matter being present in the universe, and once again the theory was modified so that it would agree with observation. It also turns out that this dark matter is not just a tiny adjustment but rather that most of the matter in the universe is dark matter. So in the next rendition of the standard model, most matter exists in some unknown form but is necessary if we wish to make the model fit with observation. Once again, dark matter is not a prediction of the early standard model. Rather, it is an adjustment that must be made to the model.

In 1998, to everyone's amazement, it was observed that the expansion of the universe appeared to be accelerating. I suppose younger folks have been taught this all along in school. But for those of us who went to school before the 1990s, this was a big surprise. The Nobel Prize for this discovery was awarded to Saul Perlmutter, Adam Riess, and Brian Schmidt. Their findings appeared to be in direct defiance of the standard model, as it was understood that the gravity of all the mass in the universe should slow the rate of expansion with time. Yet two independent teams observed that the rate at which galaxies and other galactic bodies moved away from each other actually appeared to be accelerating. Scientists therefore introduced yet another modification with dark energy (this is different than dark matter). This dark energy serves as a kind of negative gravity that forces objects away from each other and (at least in one form) is similar to the cosmological constant first introduced by Einstein—the same constant that Einstein allegedly referred to as his biggest blunder. It turns out that the cosmological constant once again became fashionable in science. Patches on, patches off, and patches back on again. With the addition of dark energy, the amount of visible matter in the universe decreases to only 5 percent. That means 95 percent of the universe exists in some combination of dark matter and dark energy. Two entities that scientists have no real idea as to what they physically are.

This big bang theory is heralded as one of the great theories, achievements, and accomplishments of human thought. But in reality, it has required a series of modifications to an initially very simple concept. The original idea was simply that the universe was smaller and smaller as you went further into the past. But along the way, scientists added an initial period of extreme expansion (inflation) and the idea that much of the universe is composed of a form of matter we do not truly understand (dark matter). And the expansion rate of the universe actually appears to be accelerating (dark energy). And, not to be overly repetitive, but none of these findings were predicted by the early standard model.

A couple of other things to note. At the time inflation was proposed, there was no strong understanding as to why the rapid expansion occurred. Over time, it has become generally believed that inflation was due to quantum-level events. However, as this has become better understood, we have learned that these quantum effects also suggest that many, many universes have come into existence and will continue to do so. This leads to the idea of the multiverse. The idea is that there is not just one universe but an array of universes with vastly different properties. This is different than the idea of a parallel universe, which is driven by the notion that all possible events occur, just in different universes.

But back to the multiverse. The question I would ask is, Why do we live in this particular universe? Out of all the universes that may be out there, why are we in this one? One explanation for this uses a variation of the anthropic principle, the idea that since we are here, the universe must have developed in a way that was compatible for human existence. But couldn't life form in a different way in a different universe? Most likely, yes. So we are back to the question of why we are in this universe with this form of life. We can go around the circle, but the ultimate question remains the same. Out of many universes, why are we in this one?

Now remember that the inflation theory was developed to avoid having to stipulate any special conditions to the early universe. No matter how time, distance, and mass sprung into existence, we would end up with a universe exactly as we observe it today. The whole purpose of inflation is to be able to say that whatever the big bang represents, it could begin in any random way and we would end up where we are today. But now we realize that inflation comes with the condition of any number of multiverses, all with different properties. If you think about this, you begin to see a problem. The inflation model does not really solve what it set out to solve. I can just as easily bypass the entire idea of inflation and say that we live in a large universe that just happened to start out with homogeneous and isotropic properties within the local area we reside in. And my justification could be that those are the conditions that were favorable for the development of human life within the portion of the universe that we experience. You see, I can make the same case that needs to be made with the multiverse model and completely avoid inflation. So what did inflation really accomplish? What problem did it solve? Remember, inflation is not predicted by the standard model; it is simply a fabrication made to avoid the necessity of predetermined conditions to the early universe. But now it appears, even with the inflation model, that our existence is based on the preset conditions of this particular universe!

Lastly, and for me personally the final straw, let's look into the age of the universe. Remember, the standard model all started with the idea of running time backward and picturing a universe that became smaller and smaller. The early standard model simply said that if we know the rate at

A BRIEF HISTORY OF THE STANDARD MODEL

which the universe appears to be expanding and run time backward at the same constant rate, we will determine the age of the universe at the time when all of it existed within the same tiny space (singularity). Since then, we have modified this idea significantly with inflation and with changing rates of expansion during different periods of the universe's evolution. Certainly, this early standard model prediction is far too simplistic and can no longer be considered a valid or accurate way to determine the age of the universe.

Guess what? That early standard model gives the same spot-on prediction of the age of the universe as what is observed. That's right, the early standard model suggests a universe that is approximately fourteen billion years old! How can that be?

Based on our more modern standard model, the universe expanded rapidly (inflation), then the rate of expansion slowed, and then—around five billion years ago—it started accelerating, and in the future it will be expanding faster and faster with time. We just happen to live in that very special moment when all these average out to be the same as the Hubble prediction. It is pure coincidence. That's right, coincidence would have it that we live in a very special time in history of the universe. But hold on. I thought this was the very thing that scientists hate. Are you telling me that the only way the modern standard model can be correct is if we impose this restrictive condition that we just happen to live in a special moment in time? Yes, that is what cosmologists today are telling you. If you lived in an earlier time or a later time, you would observe something different from the Hubble prediction, but we all just happen to live at the crossover time when observation coincidentally matches the Hubble prediction.

Even if you invoke the anthropic principle and say that there is only a small window in the development of the universe in which human life can exist, the odds are still crazy that within all those allowable times, we just happen to live in this very special moment. Something seems terribly wrong! Scientists go out of their way to suggest inflation because they do not like special conditions. But now scientists somehow are willing to say, "Oh well, it had to happen at some point, and I guess we just live in that very special moment."

What a coincidence! And there lies the problem.

30

THE SCALE METRICS PICTORIAL OF THE UNIVERSE

Below is an illustration of how the evolution of the universe is often depicted.

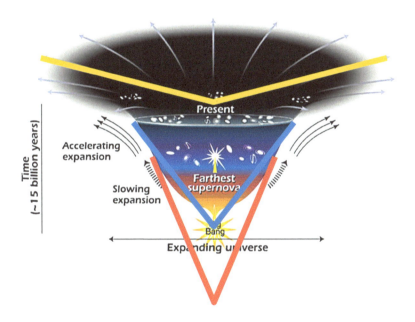

The red represents the age of the universe as predicted by Hubble in the past, the blue is the current time, and the yellow is the future. You can clearly see that the only lines that actually take you back to the beginning of the universe (the big bang) are the blue lines representing the current expansion rate of the universe. This is the special moment where Hubble actually agrees with observation. In the future it will be expanding faster, making for what appears to be a younger-than-observed universe. In the past it was expanding slower, representing an older universe. We just happen to be in the right spot!

Can scale metrics help us out of this? If scale metrics is a good model, then it, too, must be held accountable, and it must be able to address the same criticisms that I have raised against the standard model. It should be able to make valid predictions on what the universe looks like, what it looked like in the past, and how it might appear in the future. And part of the evaluation of scale metrics should be based upon whether its model, in the simplest form, agrees with observation or whether it, too, will need modifications and adjustments to reconcile the model with observation.

In pursuing the scale metrics model, I would ask the following two questions. Why is it that the Hubble prediction for the age of the universe works? And since it does work for us, why shouldn't we expect that it would work for any age of the universe? If this were the case, it would certainly get rid of the "we live in a special moment" problem.

So what might this look like?

For starters, I really dislike the earlier illustration because it suggests that the universe is expanding into preexisting space. That is not really its intent, but it sure looks that way to anyone viewing the illustration. The situation is quite different using scale metrics, where you are free to start with a square of any size. In scale metrics, the size of the square you select has nothing to do with the size of the universe. Nor is the size of the universe in any way dependent upon your

THE SCALE METRICS PICTORIAL OF THE UNIVERSE

square getting larger with time. Space is defined only by the number of free energimes present within your square. At the "beginning," your square is a complete void. The number of bound energimes that initially exist within your square is the total number of energimes that your universe will ever come in contact with. As these bound energimes phase change to free energimes, you experience a larger universe. When viewed in this manner, it becomes clear that we must use a different type of pictorial to represent the evolution of the universe using scale metrics.

An often-asked question is, What lies outside the square of the universe? There are several ways to look at this:

1) It doesn't matter; or, in mathematical terms, it is undefined. You need not concern yourself with what is outside your square because your square *is* your entire universe. Another way to see it is that the size of your square makes no difference. You can pick a large square, and I can pick a small square. In fact, if you want, you can pick an infinitely large square and I can imagine my square as a singularity. The size of the square has no bearing on the space in the universe and therefore is not a parameter that has any meaning.

2) It is simply more of the same. What your square defines is every part of the universe that can come into contact with itself from the beginning of time through its heat death. As such, by definition you are located at the center of your square (which is not the same as the center of the universe). Those in other positions define their squares in different locations—perhaps adjacent to yours. In this way the universe goes on forever, but your square only defines the portion of that universe that you could ever interact with, assuming you were present from the beginning of time all the way to the end.

In either case, the scale metrics pictorial of the universe is simply a square with an increasingly dense mesh within it as time goes by. The mesh (metric) becomes more dense because more free energimes are present. It is that simple!

So how are we to interpret the evolution of the universe using the scale metrics pictorial? Let's look into the experiences of three different individuals.

31

A STORY OF THREE JOURNEYS

This is a story about three individuals. Each brings his or her own unique perspective on the evolution of the universe.

First we meet Professor Now. She is an expert in the fields of relativity and cosmology and has focused her career on the evolution of the universe. Like most of us, she has spent her entire life on Earth. One day, as she is walking along the grounds of the university, a rugged-looking stranger appears to her and introduces himself as Aged. The two begin to talk, and Professor Now becomes captivated by an almost unbelievable story.

Aged claims that he has personally experienced the full evolution of the universe firsthand. Aged states he has been along for the entire ride and has witnessed everything from the first moment of time to the present. Despite claiming to have had this wild experience, Aged is also somewhat sheltered. He has lived his entire life in a very small region in space. Aged brings a unique perspective from across time but from a single position.

Professor Now, not wanting to be rude, politely asks Aged how old he is. He replies, "I think I am fourteen billion years old, but I may have lost count of a couple hundred million years somewhere along the way." Professor Now is thrilled to learn that Aged confirms the universe to be roughly fourteen billion years old. This is great stuff! Aged may be a tremendous resource in providing confirmation of the standard model.

Excited to learn more, Professor Now asks Aged all about the inflationary epoch. She also asks how he survived the intensely hot temperature of the early universe and whether the expansion rate of the universe is increasing, decreasing, or constant. In fact, Professor Now is asking questions so fast that Aged does not have time to respond. Yet Aged listens carefully, and when it becomes his turn to speak, he pauses thoughtfully. Professor Now apologizes for such rapid-fire questions. Aged assures her that he heard all her questions, he is just somewhat puzzled on how to respond. Aged begins by stating that he does not recall the earliest moments of the universe to have undergone a rapid rate of expansion. Certainly nothing faster than the speed of light as described by Professor Now. As for the temperature, as far back as he can remember, the average temperature of the universe has been a constant 2.7K.

Professor Now is beginning to think that Aged has not really been around since the beginning, that he must just be a deranged old man. But before she can walk away, Aged continues on. He states that as far as he can tell, the expansion rate of the universe has been constant too. With every doubling of the age of the universe, its scale has also doubled. This has been his observation since the very first moment. He reiterates: no rapid expansion, no initially hot temperatures, and no changes to the rate of expansion. Professor Now states that he must certainly be mistaken. Aged smiles and reminds her that he has actually lived through the past fourteen billion years.

It appears nothing is going to get accomplished here, so they are just about to part ways when a third individual appears. This person is a very young-looking woman who

A STORY OF THREE JOURNEYS

also claims that she has been around since the beginning of time. But she has a much different perspective then either Professor Now or Aged. She has been traveling at nearly the speed of light since the birth of the universe. She says honestly that she is a very curious individual and travels at nearly the speed of light but actually stops often to look back at where she has come from. It just so happens that she is on one of her scheduled stops right now.

She introduces herself as Traveler. She listens to what Professor Now and Aged have been saying and the apparent inconsistency in their respective versions of the universe and its evolution. Traveler states that she has had time to think about many things as she has journeyed across the universe. She tells Professor Now and Aged that there may be an explanation as to why they both see the age of the universe to be the same but strongly disagree on the fundamental physical properties of the early universe and its subsequent evolution. The three all decide to discuss this further over lunch.

It turns out that Traveler has come across an interesting interpretation of the universe. She takes out her napkin and says, "What if my square napkin represents the universe, and with time it fills up with points (or segments) that define space?" She goes on to ask, "What would happen if you thought of all that space as also denoting time?" She then suggests that they consider how the universe changes as one travels through it while she draws a line down the center of the napkin, dividing it in half. She continues by stating that when she first started out, she traveled at nearly the speed of light from the center of the napkin outward, but as she did this, the square continued to divide. She then divides the napkin into quarters. She goes on to say that her first interval of travel was from the center across a quarter of the square; with the next instant of time the square divided into

FROM FALLING APPLES TO THE UNIVERSE

eighths, and she traveled to the edge of the next section. She continued this cycle four times. Then she says, "Now, let's stop, look back, and ask the question: How far have I traveled?" (Technically she needs to conduct some calculations to account for her relativistic motion, but she is well versed in the theory of special relativity, so that is no problem.)

When trying to determine where she had come from, she knew that she had traveled four units of time, which at nearly the speed of light would have been be four units of distance. However, since her proper metric (the distance between the intervals she had drawn on the napkin) had been changing the entire time, she realized that she would interpret this distance based entirely on her current metric. In other words, she would envision that she had traveled four units of distance as defined by her current metric. She further theorized that all the larger sections that she had moved through previously where now "back filled" with energimes that are present in the current time (metric), even though they were not present at an earlier time when she moved through those areas of the universe.

Professor Now jumps in and states that this is an interesting perspective but that the standard model is well aware of the difference between the diameter of the visible universe and its much larger actual diameter. "This is all well-understood physics," proclaims Professor Now.

Traveler smiles and asks if she may continue. "What you must realize," states Traveler, "is that when I observe the universe using my current metric, I am not only fooled as to where my actual starting point is located, but I also have a misconception as to how much time exists!" Time is a measure of the number of instances in which the square has been subdivided. Traveler, at any time, looking back to where she came from, does not realize the full extent of time that has passed. "You see, the best I can tell, the universe has experienced much more time than what we can observe," proclaims Traveler. "The fourteen billion years that you state is the age of the universe, that Aged agrees is the age of the universe, and that I would agree upon (if I had not been traveling at close to the speed of light much of my life) is all based on our perspective looking along a single dimension and applying our current metric of time, distance, and mass."

At this point Aged jumps in. "Wait a minute, I have been here for the entire duration of the universe. Whatever its age is, I would certainly know—I have been here!" Traveler explains that the current metric (the one that Professor Now, Aged, and she are all presently in) does not hold as a constant across the entire evolution of the universe. In fact, it is constantly changing as more space is created on her napkin (her idea of the square of the universe). Just as you only see a portion of the total distance of the universe, you also only experience a portion of the total time of the universe (even if you have been around since the beginning).

Think of it this way: It doesn't matter whether you have been waiting your whole life for a light signal from across the universe to reach you (Aged) or whether you have only been around recently (Professor Now). You only have to be present at the time the light arrives at your location, and when it arrives, your perception of time will be determined by the local metric that you are in at that moment. Therefore, there is a difference between what I see and what I believe my actual experience to have been. What Professor Now sees and what Aged experienced are two different paths.

Very interesting, but both Professor Now and Aged ask in near unison: "What about our different perceptions regarding the expansion rate and the temperature of the universe?"

Stop and think about this for a moment. Aged has been experiencing a changing metric his entire life, yet he is not aware of this change. Think of the scale of a map. Nothing changes to the relative positions on a map if I only change the scale of the map, just as nothing happens to my physical surroundings as my scale changes when I fall through a gravitating metric. So Aged is unaware of this very real changing metric. This is why Professor Now and Aged agree on the age of the universe. But when it comes to how the universe evolved, Professor Now has only the current metric available to her, and she applies that constant metric across all of time. That is where the inconsistency comes from. Traveler goes on to state, "One of you is applying a constant metric to all of time, while the other actually lived through a changing metric with time." This results in the different physical properties of the universe that Aged directly experienced and that Professor Now perceives to be true through the use of a constant metric.

Professor Now and Aged are interested, but they are going to need some more hard evidence to be fully convinced. Traveler admits that this is as far as her knowledge can take us, but on her journey across the universe she came across a book titled *From Falling Apples to the Universe*. She suggests we all read on from chapter 32.

32

UNDERSTANDING YOUR SQUARE OF THE UNIVERSE

In this chapter I would like to further develop the idea that you can visualize the universe as a square. I have stated repeatedly that you can pick any size square you want. But does it need to be a square? After all, we generally think of the universe as a sphere. Scale metrics suggests that the third dimension is defined by a segment, so that would suggest we should look at a cross section of the universe, which would then be a circle. Isn't a square a somewhat odd geometry for envisioning the universe?

These are all good thoughts. However, the fact that our minds immediately want to think this way is also evidence of how hard it is to truly understand the concept of space. The reality of scale metrics is that there is no square or circle without the presence of free energimes. I have chosen a square simply because it is easier to incorporate the concept of a grid within a square. But the reality is that within scale metrics *there is little difference between a circle and a square.*

Take a square and place a grid within it. The grid represents the energimes that have been emitted as free space. Determining distance is simply a matter of counting the number of free energimes along any one dimension. So count the energimes (grid intersections) as you move from the center of the square to the outer edge. Now count the number of energimes as you go from the center to the corner of the square. Both numbers are the same. In both cases, for the diagram below, you move four units of space to get from the center to the outer surface of the square, just as the radius of a circle is the same to all points on the circumference. This is confusing only if you allow the idea of preexisting space to exist. That is, your mind wants to accept the idea that a square or a circle can even exist in the absence of any free energimes.

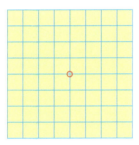

Think of it this way: When using GR, you must train your mind to accept the notion that space and space-time themselves are curved—that is, that they are not curves drawn out in flat Euclidean space but rather that the actual fabric of space and space-time is itself curved. In scale metrics we only deal with Euclidean space, but having said that, you must train your mind to accept that in the absence of any free energimes there is no such thing as space.

So what does your square of the universe really represent? It denotes the space that will be present (number of free energimes) after all bound energimes have phase changed to free energimes. At the birth of the universe, this square is a complete void. As the universe evolves, free energimes phase change until all energimes have transformed to the free state. When this occurs (heat death), the diameter across the universe will be 3.98×10^{65} energimes.

UNDERSTANDING YOUR SQUARE OF THE UNIVERSE

3.98 × 10⁶⁵ energimes

However, we will never fully experience the entire square. If you begin moving at the speed of light at the birth of the universe and travel until its heat death, you will only make it halfway across before all bound energimes have phase changed to free energimes. That is, there will be an equal number of energimes emitted in the opposite direction of which you are traveling. At the instant of heat death, there will be a total of 1.99×10^{65} energimes along your path.

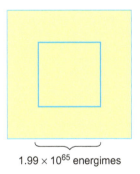

1.99 × 10⁶⁵ energimes

Next, we realize that as we travel across distance and through time that we actually only encounter half those free energimes along the way. That is because half the energimes present at heat death will have been emitted along the path after we pass through that region of space. Remember Traveler's experience: we end up at the same location (the outer edge of the square), but we just encounter fewer energimes along the way. This means that if we were observers in real time traveling through the history of the universe, we would encounter 9.95×10^{64} energimes along our path from birth to heat death. The lighter shade below is there just to remind us that we do not experience the full concentration of energimes in the universe.

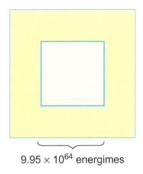

9.95×10^{64} energimes

The distance we encounter, 9.95×10^{64} energimes, is equal to the value of the Planck constant. (See section 3 for a more complete analysis of the above and on the relationship between the Planck constant and distance.)

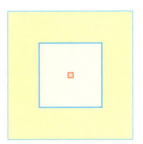

We are not yet at heat death, so what we currently see is even much less of the universe. The red square represents the present. Realize that whatever is happening within the red square is also happening everywhere within the larger squares, it is just that we cannot see everything occurring.

UNDERSTANDING YOUR SQUARE OF THE UNIVERSE

The red square represents a path that is 1/137 that of the heat death. Hopefully you recognize this as the fine-structure number!

Finally, we also know that the universe is not just empty space (free energimes). From our understanding of gravitation (section 1), we know that inconsistencies exist on very small scales in the distribution of free energimes. With time, this allows concentrated clusters to form that coalesce into structured matter and eventually form galaxies under the influence of gravitation. These galaxies exist everywhere within space in both the red square and throughout the larger squares. The universe is homogeneous and isotropic throughout.

The point I am making is that everything occurs within your square of the universe. The universe does not expand with time as observed through the standard model. There are simply more free energimes phase changing and creating more space within your square of the universe. Yes, with time you see more of space, but that is an entirely different phenomenon than the expansion or stretching of space with time. Let's explore this further in the next chapter.

33

AN EXPANDING AND COOLING UNIVERSE?

The standard model suggests that the universe has stretched with time. This is how the standard model avoids the problem of there being no space outside our space. The universe does not move outward into other preexisting space. Rather, the space available to us simply stretches with time. Technically, what we observe as stretching is the result of a change in properties and geometry to the metric of GR. Nonetheless, we observe this as a stretching in the size of the universe. This is often demonstrated with a stretching rubber band. If you put two marks on a rubber band and stretch it, the marks become further away, but they never actually change their physical position on the rubber band. In a fuller three-dimensional example, this is often modeled using a loaf of raisin bread as the dough rises in a warm oven. The individual raisins move outward but do not change their positions within the dough. The point is that you do not need more dough (space) to separate the raisins. You simply let the dough you have stretch outward, and the raisins within it will also appear to move apart.

The universe is observed to be stretching; this suggests that the universe has been cooling with time. One way to demonstrate this is with the wavelength of light. Temperature is

FROM FALLING APPLES TO THE UNIVERSE

related to the wavelength of the radiation emitted by a black-body, and the early universe is believed to be very closely related to that of a blackbody. Higher-energy radiation has shorter wavelengths, which appear blue in color, and lower-energy radiation has longer wavelengths, which appear red in color. The idea is that when the CMBR was actually emitted billions of year ago, its wavelength represented a very hot high-energy environment. As this radiation moved across space and time, the fabric of the universe stretched (just like the rubber band or the raisin bread), and when the light reaches our eyes billions of years later, is it redshifted, representing a much cooler temperature. This means when scientists claim that by looking far away they can see into the past, what they are really saying is that they can see the past from a light source that has been altered during its passage through space and time.

Scale metrics suggests a different approach. Perhaps when we look far away, we are actually seeing a light source as it was in the past! The size of the universe is determined only by the number of free energimes that exist. There is no mechanism within scale metrics that would suggest that the universe is stretching; it simply contains more of the stuff (free energimes) that define space. In scale metrics, whether you stretch your square outward with time completely misses the point. It is not the size of your square that matters, only the number of free energimes that exist. Therefore, there is also no mechanism for the universe to cool. Think of it this way: If I expand the volume of a container with a fixed number of molecules, the temperature drops. This is easily demonstrated in the laboratory and is called adiabatic expansion. These means that no energy was added to the system and that therefore as the system expands it will naturally cool. This is another way of looking at the cooling predicted by the standard model. The universe came into existence in an

AN EXPANDING AND COOLING UNIVERSE?

instant and has been expanding ever since and therefore has also been cooling.

But in the scale metrics model, if I add more particles to my square, it has no impact on the temperature. This is called an isothermal process, where temperature stays constant because more energy (or more particles) are added to the system. Think of a piece of dry ice placed inside a tied-off but uninflated balloon. The balloon expands as the dry ice undergoes sublimation and phase changes from a solid (simulating bound energimes) to a gas (simulating free energimes). This does not result in any change in temperature. The difference between scale metrics and this example is that in scale metrics the size of the square stays constant, and as you add more free energimes, a change in scale occurs that is responsible for the increasing space. Think again about the example of the mesh screen behind the windowpane.

Now some readers are about to say, "OK, I have been as patient as possible. Scale metrics may be an interesting model, but we already have solid evidence that light is redshifted as it moves across the universe. The universe has clearly cooled with time!"

I hear you loud and clear, but just hold on a moment. All we really know from observation is that light emitted long ago appears to us with a wavelength that lies more toward the red side of the spectrum. The standard model's interpretation is that light (similar in wavelength to what would be emitted today) was emitted long ago and has redshifted during its journey across space and time. This suggests that I am using my current metric and applying it to a time long, long ago. There is no assurance that this is the correct interpretation.

Scale metrics simply states that the light from billions of years ago was actually emitted with less energy because it was emitted under a different metric. That is, the light was already redshifted at the time it was emitted billions of years ago. There is no redshift due to stretching and no cooling of the universe. There are simply more free energimes, resulting in a changing metric across time—an isothermal process rather than an adiabatic expansion.

34

A LOCAL PERCEPTION OF SPACE

Hopefully you will come to see in this chapter that what is seen in the present may not truly represent what occurred in the past. Remember the example provided by Traveler. What she imagined was not what had actually occurred.

Scale metrics proposes that the phase change from bound to free energimes occurs at a constant rate from the perspective defined by your square of the universe. You can think of your local observation of distance, mass, and time as your proper scale. This proper scale defines what you measure as the energime mass, distance, and time. Collectively, it also serves to define the mesh, or grid, that is present within your square of the universe.

Remember the part from section 1 where the coordinate scale was the one far, far away and absent of any gravitational influence. This was the coordinate scale in static time when only your position changed. In section 2 we are discovering that your coordinate scale when your position is static and time is allowed to change is the scale that existed long, long ago at the beginning of time. It is the scale defined by your square of the universe, and it is constant throughout all ages.

Scale metrics suggests that the rate of the phase change is constant when measured from the coordinate scale. This is the simplest possible scenario. The model doesn't say that this is the only possibility but rather that it is the simplest

FROM FALLING APPLES TO THE UNIVERSE

assumption and therefore a good place to start. The rate of phase change also measures the passage of time as well as distance and the quantity of mass within the universe. This suggests that as coordinate time goes by (which runs at a constant rate), the diameter of the universe will increase but at an ever-slower rate as measured from the coordinate scale. Why? Because the size of the universe can be determined from a two-dimensional framework (along with a segment representing the third dimension). Remember the example of the forest that defines three-dimensional space using only a two-dimensional count of the number of trees. Area is determined as the diameter squared. Since the same number of free energimes is emitted with a constant passage of time, and since they must be distributed across two dimensions, the actual diameter of the universe will increase more slowly than the rate at which free energimes are emitted.

Stated mathematically, if I double the free energimes, the diameter increases by the square root of two, or 1.41. When I quadruple the number of free energimes, the diameter increases by two. If I increase the number of free energimes tenfold, the diameter increases by 3.16. You easily see that the rate of size increase slows down as more bound energimes are phase changed to free energimes.

But is this what you would actually see if you witnessed the universe unfold in real time? That is, rather than having to look far off into the past to understand prior events, what if we had firsthand knowledge? This is exactly the experience Aged claims to have. And, from his perspective, all the observations made in real time are from the proper scale. And this proper scale changes with time as the grid within your square of the universe becomes more dense with more free energimes.

We have already shown that the local observer is not aware of the full passage of time that has occurred. The clock in the

A LOCAL PERCEPTION OF SPACE

possession of the local observer registers less time than that which defines the rate change from bound to free energimes as defined from the coordinate scale. This relationship also turns out to be a square-root function. As coordinate time doubles, an observer's proper time changes by the square root of two, or 1.41. If you quadruple the coordinate time, the observed time that passes is doubled. And after a tenfold increase in coordinate time, the proper time that has passed is increased by only 3.16. A pattern is developing. The passage of time, as experienced in real time (the entire sequence of "present" moments), is delayed in the same way that the change in the diameter of the universe slows with coordinate time. When these two effects are combined, the result is a local observer who always measures a constant rate of increase in the diameter of the universe. (Note that I do not refer to this as an expansion, because the diameter did not stretch; there are simply more free energimes present.) This means that the observer, who actually experiences the evolution of the universe in real time, will always see a doubling in the scale of the universe with a doubling in the age of the universe.

This should not be confused with the observations from 1998 that suggest that the universe is expanding at an ever-increasing rate. Professor Now, as well as all of us, has only her current metric from which she applies this constant metric across all of time to determine her observations. The observer in real time (Aged) makes observations from the proper scale, which is constantly changing. Therefore Aged will also experience things much differently than an observer using either the coordinate scale (which is a constant) or our current metric (which we perceive to be a constant). It appears that to truly understand the universe, we must be very careful in clearly defining the metric that is being used. Further, we must understand the very nature of the metric itself.

35

A BIG BANG?

We have already gone over the standard model, but let's look a bit deeper into the first moment of the big bang, that famous event that instantly occurred from an initial singularity. The suggested proof of this is the CMBR. This is the holy grail of the big bang theory. However, if you go back to the 1960s and '70s, there were actually two competing theories: the big bang and the steady state theories. The steady state theory, proposed by Fred Hoyle, essentially said that the universe is constant over time (that it had no formal beginning) and looks the same in every direction. As existing matter continues to move outward, new matter is created to fill the void left by the other matter that is constantly moving away—hence, a steady state for the universe.

The big bang theory states that everything comes from an initial singularity that underwent a change (for unknown reasons) that created the universe with a bang. That there is a formal birth to the universe in which an initial singularity bursts into existence all at once, and then everything unfolds from that instant in time. The term "big bang" was actually a derogatory quip made by Hoyle, something along the lines of, "Oh, I suppose the universe just started with a big bang." But the discovery of the CMBR was viewed as evidence for a hot, dense, compact universe that emitted radiation related to its temperature and that this light stretched and cooled for fourteen billion years to what we see today. Having traveled across the universe, the radiation today represents a thermal signature of right around 2.7K. And this temperature is the same in every direction. There it is—proof of a big bang!

But what would a different model suggest? Scale metrics does not support the big bang. All the energy of the universe was not put into play in an instant but is being introduced throughout the ages as bound energimes (with no mass, distance, or time properties) phase change to free energimes (with mass, distance, and time properties). This suggests some similarities with the steady state model.

So what does scale metrics suggest? It brings these two major notions of the universe together, that of the big bang and the steady state. In scale metrics the universe has a formal beginning (similar to the notion of the big bang). There is an initial singularity, but this singularity does not burst into existence in a single moment. That is, not all elements of the universe are put into play at that instant. Rather, the presence of distance, time, and mass are all introduced as bound energimes (existing as singularities) phase change to free energimes. This occurs at a steady pace throughout the ages, resulting in an ever-changing metric. So perhaps what we really have is a beginning to the universe with a steady release of space over time, an ongoing change to the metric of space fueled by the time-delayed phase change of bound energimes to free energimes. Or, more simply stated:

A steady bang.

36

MATTER...IT'S NO SIMPLE THING

Remember in chapter 11 when we were trying to determine how nature might define the concept of unity? During that process, we learned about the inverse fine-structure constant and its curious value of roughly 137. Scientists have been puzzled by the meaning of 137 for decades. What is the significance of this number? In the big bang model, the inverse fine-structure constant does not change with the passage of time. Even for those theorists who suggest there could be a change, it is very modest.

So why this weird number? Well, it is only weird when we insist on giving it significance as a universal constant. Let's look at this from a different perspective. If we allow the inverse fine-structure constant to change with time, then it simply becomes a marker in time with its current value representing our current age. Since there is nothing significant about the present moment (at least there should not be), it follows that there would be nothing significant about 137 either.

We may simply be looking at a perception problem. Scientists view this as a universal constant, but what if it is actually changing with time? The question is not about the significance of 137 but rather what in the universe determines the value of the inverse fine-structure number. (Note that I am going to start referring to this as a number as opposed to a constant.) As the universe changes with time, the inverse fine-structure number may also change.

To better understand this, we need to look more deeply into matter.

This may seem simple: free energimes represent space and bound energimes are matter. But it is not this simple. I have made every effort throughout this book to always state that bound energimes are associated with matter, but *they are not themselves matter*. Seems like a little thing, but it isn't. Think about it: Bound energimes have no mass, distance, or time properties associated with them. They phase change from bound to free states, and it is the free energimes that represent mass, distance, and time. If bound energimes had these same qualities, absolutely nothing would be accomplished through the phase change from bound to free states.

It is a bit of a reversal from the way we typically think about things. We have been conditioned to think of objects as having mass, and then space simply becomes the entity that separates one object from another. In a pure vacuum (with all the gas molecules removed), we tend to think of space as a void (with perhaps some quantum-level activity). In scale metrics we completely flip this notion around. Space is the only entity that has the properties of distance, mass, and time. All the distance of the universe lies in space. All the time of the universe lies in space. And, yes, all the mass in the universe lies in space. Bound energimes, by definition, do not contain any of these three properties.

So what is matter? Matter is formed through a coupling between bound energimes and free energimes. Matter takes up space, has mass, and responds to the passage of time. These are all properties of free space. Yet matter is different than free space. Matter has form and structure to it that go beyond the pure essence of space. Matter requires a contribution of both bound and free energimes to establish its characteristic presence. We therefore have three unique, but related, conditions in which the universe can exist: free energimes representing space, bound energimes representing the singularities that fuel the creation of space, and matter that is a coupling of the singularities of bound energimes with the mass, distance, and time properties of free space.

The following example may help somewhat with this rather abstract notion. I have already shared with you that the square that defines my idea of the universe is a windowpane. We also have looked at the free energimes filling you space as the mesh of a screen behind the window. This provides the concept of space. Now also present (but unseen) are the bound energimes that phase change with time to increase the density of the mesh screen. So, with time, the screen becomes more compactly woven into a tighter and tighter mesh. Now I introduce matter, which requires space to "support" the bound energimes in a more complex structure. That structure manifests itself as matter.

So now to the model. Have you ever looked out your window through the screen on a rainy day? Have you seen how the individual squares within the screen capture water droplets and fill up that space with water? You can think of matter in the same way. Again, no analogy is perfect, and this one has many flaws, but it does give you a visualization that looks a lot like how space (the mesh) and bound energimes (water droplets) combine into a structure where the mesh supports the droplet and forms something that could not be held in place if it were not for the presence of both the mesh (space) and the water droplets (bound energimes). In fact, if you look at your window screen from a distance, you have your own little universe of space with matter sprinkled randomly throughout.

Let's look at this in yet another way. Take any object and place it on a surface near you. What happened to the space that was present in that location before your object was placed there? Now if space is a void, then there is not much of a conversation to be had. However, scale metrics suggests that space has mass, distance, and time as manifested in free energimes. Where did those energimes go when you placed your object? Did you displace the space just as you displaced the air molecules that were there? The answer is no. You did not displace the energimes associated with that space. The space is exactly where it was but has coupled with bound energimes to produce matter.

How does matter form? The standard model suggests that this is related to the Higgs boson, named after Peter Higgs. This particle was detected by the CERN (European Council for Nuclear Research) collider in Geneva, Switzerland. However, the Higgs boson provides little guidance on the specific mass of essential particles such as the electron, proton, and neutron. They exist, but their characteristic masses cannot be predicted by the standard model. These three particles combine in different ways to make all the atoms responsible for the matter in the universe. For example, the electron and proton combine to form the hydrogen atom that makes up the vast majority of all known matter in the universe (excluding dark matter and dark energy). If scale metrics can shed any light on why the electron, proton, and neutron exist with their specific masses, it might go a long way in helping to understand the role of matter in the universe. If we can also demonstrate how these characteristic masses contribute to propagate the electrostatic and strong nuclear forces, it will further strengthen the fundamental framework of the scale metrics model. Let's begin with the electron.

37

THE ELECTRON

To truly explore the nature of matter, we need to better understand the Compton wavelength, named after Arthur Compton. The electron, just as all particles, has what is called a Compton wavelength. It is the wavelength a photon would have if its energy were the same as the mass of the particle. So in the case of an electron, if you take the electron mass and convert that to the energy of a photon, the resulting wavelength of the photon is the electron's Compton wavelength.

This concept is important in physics and of particular interest within the model of scale metrics. When using scale metrics, something interesting occurs if we square the Compton wavelength of the electron and divide it by the inverse fine-structure number. The result: you get the mass of the electron! Several things to note: This is not a rough approximation but rather what appears to be an absolute relationship. Further, this does not happen if you use values for h and G that are defined by humans, such as treating them as unity. It only works when you use nature's value of unity as determined by equating the mass of the electron with the charge of the electron. In other words, an electron is an electron regardless of whether you are considering its gravitational properties or its electrostatic properties. If you wish, refer to chapter 12 for a refresher.

What might this mean? I have suggested that matter is formed through a coupling of bound and free energimes. One way this might occur is if there is an oscillation between bound and free states—in other words, an ongoing transition from bound to free and then back to bound again. If you take the bound energimes associated with an electron (which by definition have no mass) and phase change them to free energimes, you will then have a measure of the electron's mass based upon the electron's relationship with the free energimes within its proximity. Free energimes must also phase change back to bound energimes, and so this oscillation continues. If this were to occur, it would be fairly easy to see that the electron's mass would be related to its Compton wavelength. Further, if you consider the Compton wavelength squared, you have a measure of the number of free energimes present within the "reach" of the electron in free space. This is a great start, but where does the inverse fine-structure number come into play?

If there were other electrons (or any forms of matter) in the area, they would be in competition with our electron for the same available free energimes. This suggests that the inverse fine-structure number might simply be a ratio of bound to free energimes present at any given age of the universe. It represents the degree to which an electron does not have full access to all the free energimes within its reach due to the presence of other matter that encroaches within the same space.

Think of it this way: You are at a large community picnic on a rather windy day. Each of the picnic tables has a pile of napkins nicely stacked (bound state). The wind is swirling and blows the napkins all over the air (free state) not just from your table but from all the tables. Your job is to collect the napkins, but you can only reach so far before they completely blow away. You grab what you can but cannot get

THE ELECTRON

all the napkins in your reach because other people are also grabbing at napkins within your space to replace the stacks that blew off their tables. The wind temporarily dies down, and all the people get their stacks of napkins back. Then the whole process starts over again, starting with nicely stacked napkins and then another competition for the airborne napkins within your reach. And on and on it goes. Although I suppose in real life folks would eventually just place a weight on their napkins to protect them.

Think of a napkin as an electron. It goes from a bound state to a free state and then back to being bound. You can only grab the napkins within your reach (Compton wavelength squared), but even then others are grabbing at the same napkins. If the picnic tables are spread out well, then there is no competition for available napkins, but the closer the picnic tables are to each other, the greater the competition for free napkins within your reach. So the inverse fine-structure number is a measure of how many electrons are in the area as related to the number of free energimes that define the space for that area.

Let me try one additional way of explaining this. I don't think this is as prevalent as it was when I was a kid, but it used to be that during holiday parades most of the floats would have a person onboard throwing candy out to the side of the street. I think for safety reasons this is not so common anymore. But when I was growing up, this was half the fun of the parade. As the candy was thrown out toward the curb, all the children would scurry about trying to scoop up as much of it as they could. So what determined how much candy you could get? Certainly your reach came into play. Older kids with longer arms were better able to get to the candy. The other factor? The number of kids in the area. If you were lucky enough to be viewing the parade in an area that had more adults than children, then more of the candy was available for you to scoop up. If there were more kids around, the competition for the available candy was intense. It was the same amount of candy, but more arms were reaching out and grabbing for it.

That is basically how the number 137 works when viewed within the scale metrics model. It is a representation of how much competition there is for the free energimes. As bound energimes phase change to space, the relationship between bound energimes and free energimes constantly changes as the universe ages.

Therefore, as I have suggested, the inverse fine-structure number may not be a constant at all. It may simply be a number, 137, that represents the ratio between bound and free energimes at the present time. It constantly changes with the evolution of the universe as bound energimes phase change to free energimes. At the beginning of time, there were only bound energimes and no space. As time goes by, more space is created, and the space between electrons (and other forms of matter) increases. At the end of days, there will be only space.

So the fine-structure number suggests that 1/137 of the energimes along a single dimension, at this mark in time, have phase changed to free energimes. It also suggests that the value of the fine-structure number ranges from zero in the beginning (no space) to one at the end (only space). Right now, it appears we live in a fairly young universe, with only 1/137 of bound energimes having undergone a phase change. We have a long way to go before all bound energimes have phase changed to space.

This all suggests the existence of a stable particle whose Compton wavelength squared and divided by 137 is equal to the particle's mass. What does that turn out to be? Exactly the mass of the electron! It further suggests that the mass of the electron is not constant throughout time but changes as the inverse fine-structure number changes. I realize that some readers will resist the notion of a changing fine-structure number or, for that matter, a changing mass to the electron. Please read on; there is more to follow in future chapters.

For now, we have a model that exactly predicts the mass of the electron. Let's see if we can do the same for the proton.

38

THE PROTON

The existence of the proton has been known for over a century based on the work of Ernest Rutherford. Today, the accepted value for the mass of the proton is 1.67×10^{-27} kg. This is 1,836 times more massive than the electron. Just as we asked for the electron, why has nature assigned this value to the proton? The standard model offers little guidance.

To help us answer this, let's explore what we know about the proton. It is a charged particle with an equal but opposite charge to that of the electron. This creates an attractive influence between protons and electrons but a repulsive influence between proton or electron pairs. The proton, along with the neutron, makes up the nucleus of an atom. The strong nuclear force is the influence that holds the nucleus together. Remember, all the positively charged protons repel each other, so in any atom with more than one proton, some type of influence is needed to keep them all packed together in the relatively small confines of the nucleus.

Using scale metrics, we have the energime (propagating gravity) and the electron (propagating both gravity and electrostatic force). And now we need to consider the proton (propagating gravity, the electrostatic influence, and the strong nuclear influence). Ideally, we want to be able to describe this influence using the same scale metrics model that governs gravitation and electrostatic influences.

However, this might appear to be a bigger challenge. If you recall, both gravitation and electrostatic influences can be described using classical equations that are very similar to each other (chapter 12). Furthermore, both of these influences extend their reach all the way to infinity.

The strong nuclear influence is quite different. It operates over a very short range and has a nearly constant influence within that range. Its influence then falls off sharply once beyond its reach. It truly can be thought of as a deep well that traps the positively charged protons within its walls.

Protons viewed as "trapped" in a deep well, unable to get out

This is a much different picture than what is portrayed by either gravitation or electrostatics. Can scale metrics accommodate this? If so, how?

When you think of the electron propagating the electrostatic influence, it becomes clear that a void in free energimes is created by the presence of the electron. This provides an environment ideal for a new influence of limited range—a new type of particle that also oscillates between bound and free states. Except in this case, as the particle phase changes back to bound energimes, there are no available free energimes to pull from within space (they have all been used to generate the electrostatic influence of the charged particle—either positive or negative). Rather, any phase change that occurs from bound to free must use those exact same energimes in the phase change from free back to bound. It is basically like a yo-yo on a string. I can throw the yo-yo out, but that same yo-yo must return. It is limited in its range by the length of its yo-yo string.

The electrostatic charge is quite different. If I throw the yo-yo representing an electron out, the influence of the electrostatic charge is not attached to a string. It flies out and continues on a path all the way to infinity. The energimes needed to phase change back to the electron are pulled from free space; they are not the same energimes that were phase changed to free space. This is what results in an oscillating wavelike disturbance in space that propagates the electrostatic influence. Depending on the phase of this wave, the electrostatic charge can produce either an attractive or repulsive influence. Since the proton carries a positive charge, we can refer to its electrostatic influence as being due to a positron (same as an electron but opposite in charge).

So now we have a new particle, a kind of self-regenerating particle (SRP) that cannot exist separate from the electrostatic influence. And what do we know about this particle?

- By definition, its Compton wavelength squared must be equal to its mass.
- This means it is not influenced by the inverse fine-structure number, and intuitively we can expect it to propagate a force 137 times stronger than the electrostatic influence.
- It operates only within the average reach of the positron's Compton wavelength; or, half its Compton wavelength squared.
- All the bound energimes that undergo a phase change to free energimes must be the same energimes that phase change back to the bound state. This is the yo-yo effect.
- Nothing "leaks out" (its range is limited), as it operates only within the void of free energimes left by the positron.

Based on what we know now, determining the mass of the proton becomes a multistep process:

1) Since the positron uses all the available space within its reach, it essentially creates a void without any free energimes. We must reestablish a background field from which to compare the strength of the strong nuclear influence. The value of this background is the electrostatic mass. We then increase the newly generated background 137 times to establish the strength of the strong nuclear influence. Therefore, we must account for 138 electrostatic masses.
2) The Compton wavelength squared for an SRP is 2.21×10^{-25} m². Compare this to the much larger area of one half the positron: 2.94×10^{-24} m². We can therefore "fill up" the additional positron space with more SRPs.
3) We must account for the electrostatic influence present within the proton. This requires one additional electrostatic mass.

When the above steps are considered, the mass of the proton comes out be 1.67×10^{-27} kg. An exact match with the known mass of the proton! If you are interested, the math is shown below:

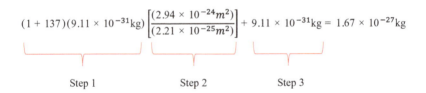

39

THE NEUTRON

So what about the neutron? While it is true that most of the matter in the universe consists of hydrogen (a single proton and electron), one cannot fully understand matter without the existence of the neutron. This is a neutrally charged particle that shares space with protons within the nucleus of many atoms. The neutron is essential for the formation of more complex atoms. If you look at the periodic table, the atoms with larger atomic numbers are partially made possible due to the presence of the neutron. That is because the neutron exhibits the strong nuclear influence but has no charge. Remember, the protons in the nucleus repel each other. The neutron acts as a buffer to some degree in keeping protons away from each other, thereby stabilizing the nucleus.

The mass of the neutron is a bit more than that of the proton, at 1.675×10^{-27} kg. Again we ask why nature has assigned this mass to the neutron. Once again, the standard model has no answer. Being neutrally charged, the neutron is easily thought of as a combination of a proton and an electron. This notion is further fueled by the observation that the neutron undergoes radioactive beta decay, which transforms the neutron into a proton, an electron, and an antineutrino. However, there are many problems with the naive interpretation. First, the sum of the proton and electron mass does not get you all the way to the neutron mass. The neutron requires the equivalent of 2.53 electron masses to achieve its mass. Second, by all accounts, the neutron is its own particle. It is not made up of a proton and an electron. Yes, it undergoes radioactive beta decay, but simply because it can decay into a proton and an electron does not mean that it is composed of them. Third, scale metrics (as well as quantum mechanics) prevents an electron and proton from sharing the exact same position. Yet if you want to create a neutral particle by combining a positive and negative charge, they must be superimposed on top of each other to exactly cancel out the influence of their respective charges. If this were to occur, the electron and positron charges would actually annihilate each other with a large expulsion of energy.

Let's see if scale metrics can offer a solution.

As with all matter, the neutron must achieve its mass through a coupling with free energimes. And, since it is a neutral particle, we must find a way to block or capture any free energimes that might otherwise escape into the greater space and manifest themselves as some type of influence.

One way to picture this is to think of a way in which the positron can be viewed as an SRP. That is, find a way to require that the specific energimes used in the positron's phase change from bound to free be those same energimes captured during the phase change from free back to bound. This would require the presence of an additional particle whose role would be to utilize all the available free energimes within the reach (Compton wavelength squared) of the positron.

This will require a particle whose Compton wavelength squared can account for its own mass along with the full mass of a charge. This is rather straightforward and turns out to be a particle with 0.755 electron masses. It is important to note that just as it is with the SRP that helped define the proton mass, there is only one mass that achieves the desired outcome. We are not able to manipulate our model with various options but have only one mass that meets the requirement of a particle that can account for its own mass and also "capture" all the mass of a charge. But, again, just as with the proton mass, we are not quite done yet. You need to make an adjustment for the difference in wavelengths between this capture particle (CP) and the wavelength of the positron, which now acts as an SRP.

A simple way of looking at this is to ask how many positrons can fit in the space provided by the Compton wavelength squared of the CP. The final step is then to add the mass of the CP, and you should then get an estimate of the additional mass of the neutron as compared with the proton.

So here are the steps:

1. You start with the mass of a charge: 9.11×10^{-31} kg.
2. You make an adjustment for the ratio of the CP Compton wavelength squared to the positron Compton wavelength squared.
3. You add the mass of the CP.

And this is what you get:

$$(9.11 \times 10^{-31} \text{kg}) \left(\frac{3.21 \times 10^{-12} m}{2.42 \times 10^{-12} m} \right)^2 + 6.88 \times 10^{-31} \text{ kg} = 2.29 \times 10^{-30} \text{ kg} = 2.51 e$$

Step 1   Step 2   Step 3

This is within 1 percent of the observed difference between the proton and neutron mass of 2.53e.

Before moving on, I do want to acknowledge that the models presented for the mass of the electron, proton, and neutron are fundamentally intended to show how matter might form through a coupling with free space. This appears to be a much different approach than that taken by the standard model, which suggests that the proton and neutron are composed of quarks. However, no matter how you approach the subject of fundamental particles, I suggest that a large gap in knowledge still exists in fully understanding the true mechanisms by which they form. Why can I say this with confidence? Because to truly understand the nature of matter requires understanding interactions all the way down to the level of the energime. In the case of mass, this means understanding interactions occurring at the scale of 10^{-73} kg. The electron mass is larger than this by a factor of 10^{42}.

For comparison, the difference in scale between an atom and the Milky Way galaxy is about 10^{31}, not even coming close to the difference between the energime and the electron. Yet you would never say that because you observed a galaxy that you then had all the information needed to understand all the complexity, forces, and influences operating within that galaxy. So it is with attempting to understand matter.

This further suggests that the weak nuclear force (the one responsible for the beta decay of the neutron) may well be the first of many influences to be discovered that play a role in how matter forms and is held together.

In any case, we are far from understanding the true nature of matter and currently have no tools for getting there. Even the strongest of proposed particle colliders will fall dramatically short of providing any meaningful attempt to understand the deepest secrets on the formation of matter. An entirely new, and at this point unknown, approach will be required.

40

A LITTLE MORE ON 137

We have now used scale metrics as a model to show how bound and free energimes might couple to form the electron, proton, and neutron. With this we have the general concept necessary to account for the full array of matter in the universe as mapped out on the periodic table. It is now time to take this information and return to our conversation on a changing value to the inverse fine-structure number, a topic I promised you we would revisit. This time I will start by discussing what exactly is meant by the term "fine structure."

The inverse fine-structure number was proposed by Arnold Sommerfeld in 1916 and is sometimes referred to as the Sommerfeld constant. It was first introduced as an extension to the Bohr model of the atom. The Bohr model works best for the simplest atom, the hydrogen atom. Hydrogen is composed of only one proton and one electron. The Bohr model treats the electron as if it were orbiting the proton, just as a planet orbits the sun. However, in the case of the hydrogen atom, the electron is held in orbit through an electrostatic influence rather than gravitation.

It turns out that the Bohr model makes some excellent predictions for the hydrogen atom. Yet in other areas, such as understanding electron spin, it does not work so well. So it is a model, one that in some cases mimics and simulates

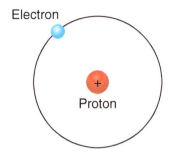

reality quite well, but it is certainly far from a complete description of the atom.

What this means is that we can use this model carefully by understanding its limits. One of the ways in which the Bohr model performs very nicely is in describing the spectrum of light emitted by hydrogen. It also serves as an excellent introduction to quantum mechanics. Often, when one hears the term "quantum mechanics," it comes with the notion of strange and hard-to-understand phenomena that occur within the tiniest scales of the microworld.

But in actuality, many of the fundamentals of quantum mechanics are very simple. In the case of the Bohr atom, it turns out that the electron orbitals can only exist at certain energy levels that correspond with multiples of the electron's wavelength. These restrictions to the allowable electron orbits define the various energy levels. The term "quantum leap," when applied to the spectrum of hydrogen, simply means that an electron must jump from one energy level to another without the ability to be in between. It is similar to a flight of stairs. Whether going up or down, you can only move from step to step and cannot stop midway between any two steps. You can take the steps one, two, three, or more at a time. The point is that you cannot stop between the steps.

A LITTLE MORE ON 137

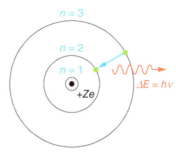

When an electron falls from a higher energy level to a lower one, it will emit light of a certain wavelength. When it falls between other orbital combinations, other wavelengths are emitted. This results in a specific spectral signature given for hydrogen as well as for other atoms. The same is true when the electron absorbs energy. An electron's ability to absorb energy is determined by its movement from a lower-energy to a higher-energy orbital.

When light is given off in a continuous spectrum, we are able to see a portion of that as visible light. It ranges from red to violet. Just outside the range of visible light on both sides of the spectrum lie infrared and ultraviolet. What we see is the continuous spectrum of visible light. When hydrogen emits light, it does so only at specific wavelengths (color bands). When hydrogen absorbs light, it pulls only from those same color bands, leaving an absence of light.

Below is a diagram of the continuous spectrum as well as the hydrogen emission and absorption spectrums. The horizontal numbers below the spectrum identify the specific wavelengths, in nanometers, for the corresponding colors of light.

Again, you can think of the bands of light in the hydrogen emission spectrum as if the electron were moving along a staircase with steps that are not all equally spaced, and it can also jump through different numbers of steps at a time, so the gaps are not equal in length.

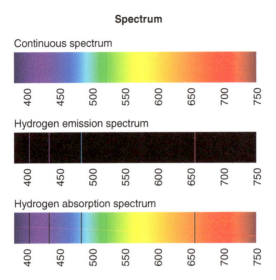

If you just focus on the red band of light in the hydrogen emission spectrum and you magnify this many times, you will find an additional hairline split in its spectral line. This tiny split is referred to as a doublet, and it defines the fine structure referred to in the inverse fine-structure number.

Sommerfeld proposed that if you loosen the requirements of a circular orbit and allow for elliptical orbits of the electron around the nucleus, these might account for the observed doublet. His idea worked, but the success of this approach was limited to this particular application. Just as with the Bohr model of the atom, it fell short in other areas, such as fully understanding the relationship between electron spin and the electron's orbital angular momentum.

In any case, Sommerfeld found that the fine structure was related to the velocity of an electron as compared with the speed of light. In fact, to achieve the observed fine structure, the speed of the electron needed to be 1/137 the speed of light. Hence the inverse fine-structure number was 137.

Today, the inverse fine-structure number has many other more-abstract interpretations. It is seen as the ratio of the

A LITTLE MORE ON 137

coupling constants between the strong nuclear influence and the electrostatic influence. This is a fancy way of saying the strong nuclear influence is 137 times stronger than the electrostatic influence. As you may recall, this is the very information we used to determine the mass of the proton.

This brings us to a very interesting possibility. What if the structures of the electron and proton (and therefore the hydrogen atom) were different in the past? This would certainly suggest a changing value to the inverse fine-structure number. And as the properties of the hydrogen atom change, so will its spectral signature. That is, the hydrogen atom might well have emitted light billions of years ago that was already redshifted at the time of its emission!

Think about the things you typically observe. Do most of them change with time? Is it really that absurd to think that a hydrogen atom that existed billions of years ago might have looked different from a hydrogen atom today? Doesn't it seem somewhat naive to think that matter existed exactly the same in the past as it does in the present?

In fact, this is a recurring discussion in modern cosmology. In one instant, the universe did not exist. An instant later—a big bang—and it sprung into existence and has simply been stretching for fourteen billion years based solely on what transpired during the "bang." Is it not just as plausible—or maybe even more so—that the mass, distance, and time of the universe have continued to change as the universe has evolved? In scale metrics this is accomplished through the ongoing and steady phase change of energimes from bound to free states. In other words, the initial singularity did not burst into its entirety in an instant. Rather, the phase change from bound energimes (singularities) to space occurs at a steady rate over time.

And let's return briefly again to section 1 and the discussion of the event horizon of black holes. Once again, the standard model suggests that on one side of the event horizon, you are free to return to a distance infinitely far away from the black hole. And then—poof—one tiny interval closer, and you are forever trapped within. Is it not plausible that the scale metrics suggestion of a smoother change in the metric that prevents light from reaching you far away may be a possible—and perhaps desirable—option to consider?

And what about the CMBR? At 380,000 years into the universe, a flash of light appears that has been traveling from that instant across the universe and has been cooling as space and space-time stretch. When we view that radiation, is it really all coming from just that single period of time? Perhaps not!

The bottom line is, scale metrics suggests a much smoother and continuous change in many properties (including the speed of light, structure of matter, value of the inverse fine-structure number, value of the metric, and strength of electrostatic and strong nuclear influences) between the first instant of the universe and the present.

In most every other field of science, the component of change is considered. However, in cosmology, scientists tend to look back as if the rules of today (our current metric) accurately apply to the past (the metric of the past). This may be convenient, but it does not ensure that it is indeed correct. Perhaps we live in an ever-changing universe that includes ongoing changes in what we thought were universal constants.

41

AN EVER-CHANGING UNIVERSE

Here is the big question for all of science. If we live in an ever-changing universe—and nothing is constant—what is the role of science? The whole idea behind the physical sciences is to find those things that are constant and learn how to apply them to better understand our world. For example, if gravitation were different every time you dropped something, there would be no benefit in attempting to quantify our observations for falling objects. If we could not do this, there would be no section 1 of this book. And without section 1, we would be at a loss to apply what we learned about gravitation to better understand the universe in section 2. The point is, for science to work, some things do need to be constant. The question is, what? To find the best answer, we need to be open minded.

We must remember that all the laws of nature are human-made and that nature may, or may not, pay any attention to them. They are laws only to the extent that they make sense to us and, based on our limited observations of the universe, seem to hold true. That is a far cry from stating that humans can define the laws of nature. Perhaps nature just is. It exists! But humans obsessed with attempting to understand and control it feel they must assign laws to what they observe. The point is, we should not get too comfortable with what we declare the laws of nature to be.

Keeping this in mind, I have suggested that the speed of light is not constant for all ages. I have further suggested that the value of the metric changes with time. I have suggested that the structure of matter changes with a changing metric. I have suggested that the inverse fine-structure number evolves with the age of the universe. So if none of the above is constant, then what is?

Scale metrics suggests that the one thing that remains constant is the rate at which things change—more specifically, the rate at which energimes phase change from the bound to free state as measured by the coordinate scale defined by your square of the universe. (And this suggests a uniform energy density to the universe resulting in a constant temperature regardless of when or where the blackbody radiation was emitted.) You might be wondering what the basis is for this claim. I would say you are asking the wrong question. I am not making a claim; I am simply suggesting a model. Let's apply the model in its very simplest form and see whether we get results that mimic and simulate actual observations.

Let's see whether we can explain redshift and predict the age of the universe. Let's see whether scale metrics predicts a universe that is homogeneous and isotropic without a need for inflation. Let's see whether our model predicts the existence of dark matter and provides an explanation for its physical existence. Let's see if our model explains the relationship between redshift and the location of Type Ia (pronounced "one 'a'") supernovae that introduced the notion of dark energy. These are the factors that provide strength and credibility to a model. The model drives the process; it is not constantly altered to bring it into agreement with observation!

42

REDSHIFT

The early standard model was actually born out of the observation of redshift, which suggested that distant galaxies are moving away from each other due to the stretching of space and space-time as the universe evolves. Yet scale metrics has no previsions to support this movement or stretching of space.

Scale metrics suggests that light from the past was emitted at a different wavelength because it was emitted by atoms with different properties. The atoms from an earlier age had different properties because the universe was governed by a different metric. We are going to be using the inverse fine-structure number often in this chapter. We know now from the Sommerfeld meaning of the inverse fine-structure number that its value is related to the atomic spectrum of light from hydrogen. Our challenge is to look at the relationships between a changing metric, a changing inverse fine-structure number, and a changing value of the electrostatic charge and see how these change the energy of light emitted by the hydrogen atom as it may have existed in the past. It sounds like a difficult task, but we have all the tools in place to accomplish this.

Let's start with a universe that is half its current age—or seven billion years old, as observed using the proper metric in real time. In other words, we are going to track this from the perspective of Aged, and therefore the universe will double in diameter from its size at seven billion years to the present. Under these conditions, the wavelength of the light emitted seven billion years ago would have also doubled as it stretched across space and time when applying the ideas of the standard model. Cosmologists use a special notation for tracking redshift, denoted with the letter z. For our example, the value of z would be one. It is essentially a measure of how much the light stretched relative to its original wavelength. When the wavelength doubles, it has a value of $z = 1$. If it triples, it is $z = 2$. When it quadruples, it has a value of $z = 3$, and so on.

Using scale metrics, at half the age of the universe, half as many free energimes would have phase changed from bound to free as compared to the present time. This would result in the inverse fine-structure number increasing by the square root of 2, providing a value of 194. For now, we are mostly interested in how a change in the inverse fine-structure number will impact the charge of the electron. Remember, charge is a bundling of energimes that is equal to the mass of the electron. The electron mass is determined by its Compton wavelength squared divided by the inverse fine-structure number.

This means that if we increase the inverse fine-structure number by 1.41, the mass of the electron will decrease by the cubed root of 1.41, or 1.12. Since the charge of the electron is directly related to its mass, it follows that the electron that existed seven billion years ago carried a charge that was weaker by a factor of 1.12. This changes the properties of the hydrogen atom, the mass of the electron and proton, the radii of the various orbitals, the velocity of the electron, and the energy-per-unit mass in each of the orbitals of the Bohr atom. These factors will all be addressed in more detail in section 3, but fortunately many of these changes balance each other out, leaving the critical factor as simply the charge of the electron. It can be shown that the velocity of the electron within any of the Bohr orbitals is proportional to the charge on the electron cubed.

So watch what happens: At half the age of the universe, the charge on the electron is decreased by 1.12. Its velocity in any orbital is decreased by 1.12 cubed, or 1.41. A decrease in orbital velocity of 1.41 increases the inverse fine-structure number from 137 to 194, maintaining the desired relationship between electron velocity and the inverse fine-structure number. Energy-per-unit mass is proportional to velocity squared, thereby decreasing this value by a factor of 1.41 squared, or 2.

Therefore, the energy per photon of light emitted by the hydrogen atom seven billion years ago was exactly half of what it would be today. Each photon emitted seven billion years ago had a wavelength that was already redshifted at the time it was emitted. It then traveled across space and time and reaches us with a wavelength exactly double what the wavelength would be for a hydrogen atom today.

Each energime within a photon emitted seven billion years ago is detected today possessing the same energy that it emitted seven billion years ago. We observe that today as a wavelength of $z = 1$. But in our case, z does not represent a stretching of the wavelength but merely the realization that under the metric that governed the universe seven billion years ago, the hydrogen atom from that time emitted light with an energy per energime that is exactly half the energy per energime seen today.

So why do scientists say that the inverse fine-structure number has not changed over the evolution of the universe? Because the standard model suggests that light given off long ago behaved much as it would today. They then conclude that the light redshifted over time and distance to what we observe today, assuming a constant value in the inverse fine-structure number of 137.

Using scale metrics, the wavelength (through our current metric) is constant and the inverse fine-structure number changes. Two very different models. What I wanted to demonstrate in this chapter is a mechanism to show how the universe long ago may have emitted light in a much different manner than what is experienced today. This is achieved without any receding of the galaxies attributable to the stretching of space and space-time.

Next up, the age of the universe.

43

AGE OF THE UNIVERSE

How old is the universe? The honest answer is that we don't know. The reason we cannot know the age of the universe is that any attempt to determine its age is dependent upon interpreting observations within the framework of an existing model such as the standard model, scale metrics, or some other model.

None of these models are absolute; therefore, they introduce potential weaknesses in their ability to definitively state the age of the universe. Anyone who tells you differently is simply buying into a particular model hook, line, and sinker. This is often what happens in the modern field of cosmology where—at least in my opinion—the scientific community comes dangerously close to treating both the standard model and GR as absolute. Yet anyone who addresses this question honestly will need to concede that we will probably never be able to say with complete confidence how old the universe is. Yet this is still a question worth pondering. Just realize that our answer is wide open to being wrong.

Using scale metrics, one might first ask how old the universe will be when it dies. This would be the interval between the universe existing entirely of bound energimes (birth) to the time when all energimes will have phase changed to free energimes (death). This is the range between fine-structure numbers of zero at the beginning and one at the end of days.

We know that we currently exist at a time with a fine-structure number of 1/137, or 0.0073. So as suggested earlier, we once again see we are on the very early end of this journey from the birth to the death of the universe.

Scale metrics claims that the distance across the universe at its heat death will be 10^{65} energimes. This means that within the square of your universe, there will be 10^{130} free energimes. However, this is not the same as the total number of free energimes.

As stated in chapter 32, even when traveling at the speed of light, the best you can accomplish is to travel halfway across the universe from beginning to heat death. And even then, you will only experience half those energimes along that path. If you need to revisit chapter 32 or would just like to think about this, please take some time to do so. We need to take a break anyway so that I can go outside and shovel the snow that has just begun to fall.

OK, I am all bundled up outside, and I have begun to shovel. The snow has just started falling and appears to be coming down at a steady rate. I have started at the top of my driveway and am shoveling a single path down to the street. As I proceed, I periodically stop and turn around to see how far I have gone. What I realize is that when I look backward, I see that I have only removed (encountered) half the snow that has fallen along my path. That is because half the total snow that has fallen along my path thus far has fallen behind me after I had already been past that point with my shovel.

If we think of the snowflakes as free energimes, it is easy to see that as I move through space I only encounter the free energimes that lie ahead of me. The energimes that are phase changed behind me, after I have already passed through those areas, have no impact on my journey. That affects both the distance I see myself having traveled and the time that I believe has gone by (as well as the amount of

AGE OF THE UNIVERSE

mass I see within the universe). Or, more simply stated, it affects the amount of space I am aware of.

Right now, if you apply this snow-shoveling story to the free energimes in scale metrics, then I have only moved 1/137 of the way down my driveway. And since I have only actually removed half the snow, I have only encountered 1/274 of all the snow that will fall along this path. Also note that by the time I get to the end of my driveway (heat death), what I have already accounted for along the small portion that I have shoveled (1/137 of the driveway) will be duplicated an additional 274 times as I work my way to the end of my driveway.

This information provides me with a strategy to determine the age of the universe. This is exciting, so enough with the shoveling. I am heading back inside to do some calculations.

Using what I have imagined from the snowstorm, I can state that the length of my driveway represents the heat-death distance of the universe. This is 9.95×10^{64} energimes. This is also related to the heat-death age. But, in our current time, I have only experienced 1/274 of 1/274 of the energimes to be phase changed during the history of the universe. So the current age of the universe is

$$\frac{9.95 \times 10^{64}}{(274)^2} = 1.33 \times 10^{60} \text{ energimes} = 4.48 \times 10^{17} \text{s} = 14.2 \text{ billion years}$$

This is consistent with the current estimate of a universe that is 13.8 billion years old!

There is a second way to determine the age of the universe using scale metrics. It will provide the same result, but it takes us through a different thought process that gives us a richer perspective on what is occurring in both the proper scale and coordinate scale.

At the current time, 1/137 of free energimes along a single dimension have phase changed. This suggests that, of all energimes present in the universe, that $(1/137)^2$ of energimes have phase changed. If we can determine the rate at which energimes undergo phase change, we can determine the age of the universe.

The phase-change frequency and wavelength of the energime (unit value of one) can be determined from Planck's constant. The energime wavelength is equal to Planck's constant, and its frequency is equal to the inverse of the Planck constant.

This comes directly from the well-established equation relating frequency to energy, stated below as

$$E = h\nu.$$

Here, E is energy, h is Planck's constant, and ν is frequency.

Therefore we can determine the observed cycles per second for energime phase change as follows:

$$\left(\frac{1}{9.95 \times 10^{64}}\right)\left(\frac{1 \text{ energime}}{3.38 \times 10^{-43} \text{ seconds}}\right) = 2.97 \times 10^{-23} \frac{\text{energimes}}{\text{second}}.$$

This is our observed frequency. However, we also know (from both chapter 32 and the snow-shoveling experience) that we only observe half the total energimes emitted. This means that the actual frequency must be twice as large to account for what we directly observe. It also suggests that the age of the universe as observed by us is only half its actual age.

We can then determine the age of the universe by the percentage of energimes that have already phase changed:

AGE OF THE UNIVERSE

$$\left(\frac{1}{137\times137}\right)\left(\frac{1}{2\times2.97\times10^{-23}}\right)\left(\frac{\text{seconds}}{\text{energime}}\right)=8.97\times10^{17}\text{ seconds.}$$

This represents the actual age of the universe, of which we only observe half, suggesting that the observed age of the universe is

$$\frac{8.97\times10^{17}}{2}\text{ s}=14.2\text{ billion years.}$$

Next up is inflation.

44

INFLATION—IS IT NEEDED?

Remember that inflation is a notion that allows the universe to start in any random way and end up looking like it does today. Distant areas of the universe are just now coming into contact with each other and have not had time to smooth out any differences from the early universe. Inflation suggests a rapid expansion that smoothed things out during the earliest moments of the universe. This is all needed because the early standard model does not predict a homogenous and isotropic universe. Inflation was introduced to bring the model into agreement with observation.

Scale metrics suggests something quite different. The total time that has passed is determined by the total number of free energimes multiplied by the unity value for time. This provides the coordinate age of the universe. In scale metrics, there is a significant difference between the coordinate age and the proper age. The proper age is related to the square root of the coordinate age. This should not be surprising, because it is the same reasoning used to demonstrate why the scale of the universe doubles with a doubling of its proper age (see chapter 34).

At 14.2 billion years, from the proper scale, we will observe 1.32×10^{60} energimes. This translates to 2.81×10^{121} energimes emitted within your square of the universe, representing 9.5×10^{78} s.

This is roughly a factor of 10^{61} more time than what we think has occurred. In other words, this is more than enough time for all regions of the universe in which we will ever interact with to have come to an equilibrium. This is a direct result of the simplest version of scale metrics, and it predicts a homogeneous and isotropic universe with no need for any corrective adjustments!

Another way to look at this is to compare the coordinate time, which is constant, to the value of the proper metric, which changes with time. At the beginning, the coordinate and proper scales where the same. As time goes by, the current observer continues to see less time pass relative to the coordinate time. At 14.2 billion years, what we measure as one second within the current proper scale was equivalent to 2.21×10^{61} seconds in the coordinate scale during the first instants of the universe.

This does not represent an expansion rate faster than the speed of light, as suggested by the standard model. If you were present at the beginning, you would have seen nothing abnormal. The universe increases in size at a constant rate, always doubling in scale with a doubling of time (as seen in real time by the proper scale). Yet it clearly provides an opportunity for the universe to reach an equilibrium so that areas seen at great distances and in any direction today will indeed be seen as homogeneous and isotropic. All without the need for a special inflationary epoch!

Next up are dark matter and dark energy (the missing mass).

45

DARK MATTER AND DARK ENERGY

Scientists know from observation that there is additional material in the universe that provides gravitation beyond the type of matter we are aware of. In other words, there is something out there in addition to atoms and molecules made of protons, neutrons, and electrons. And, as has already been noted, the standard model predicts that this unknown material (as some combination of dark matter and dark energy) makes up 95 percent of all the material that emits gravitation. So what is this other 95 percent? This is one of the big mysteries of our times.

When applying scale metrics, matter forms from a coupling of bound energimes and free energimes. Therefore there is a limit to the amount of matter that can exist. Once all of space is associated with matter, any additional bound energimes are unable to participate as matter. That is, there is no available space for the remaining bound energimes to couple with to form matter. However, the remaining bound energimes still emit a gravitational influence. They do not need to be in the form of matter to provide gravitation. Bound energimes phase change to free energimes, and this is the very essence of gravitation. Local fluctuations in the large-scale structure of the universe provide the very concentration of bound energimes that allows for the formation of galaxies. Therefore, within any galaxy, one would expect a larger concentration of bound energimes to be present.

Now some readers may say, "Wait a minute, are you claiming that all of space is consumed by matter? If so, why can I see space all around me?"

This is where nature has done us all a big favor in that the energime is not a point but rather a segment. If all the universe were compressed to a two-dimensional surface, then yes, all of space would be coupled to bound energimes in the form of matter and only matter would exist. Why is this not the case? Because the energime, as a segment, allows matter to be spread all along the energime segment. Once again, think of the forest of trees. Each tree may be affiliated with matter at only one point along its trunk (except for the special condition of an entangled particle). The rest of the tree trunk continues to define free space.

Another way of looking at this is to picture the leaves that fall to the ground in the autumn. When distributed only along two dimensions, the leaves cover all the ground. Yet during the summer months, before the leaves fall, there is all kinds of space between the leaves, which are all distributed at many different heights along the tree. So if I take a flat surface that consists solely of matter but allow it to extend upward to different positions along the segment of time, I do indeed have space that will appear between different pieces of matter.

How much matter is in the universe right now? That is fairly easy to determine. For this discussion, we do not need to worry about what we can see versus what might actually be happening. That is because the inverse fine-structure number has the same ratio for the entire universe (whether we have interacted with it or not). The total number of energimes in the universe is $(4h)^2$, or 1.58×10^{131} (chapter 32).

DARK MATTER AND DARK ENERGY

Currently, how much of this is space? This is equal to 8.43×10^{126} energimes:

$$\frac{\text{Total energimes}}{\text{Portion yet to phase change}} = \frac{1.58 \times 10^{131}}{(137)^2} = 8.43 \times 10^{126} \text{ energimes.}$$

This would mean that the universe can support 2.02×10^{84} electrons:

$$\frac{\text{Available energimes}}{\text{Energimes per electron}} = \frac{8.43 \times 10^{126}}{4.17 \times 10^{42}} = 2.02 \times 10^{84} \text{ electrons.}$$

But the universe is neutrally charged, so it would contain 1.01×10^{84} electrons and an equal number of positrons that would couple with SRPs to form protons, resulting in a universe made primarily of hydrogen. (More than 90 percent of the universe is believed to be hydrogen—so not a bad assumption.) This would mean that the total mass of the universe (if entirely hydrogen) would be 1.70×10^{57} kg. (Note this is not the mass of the observable universe but the mass of the entire universe as it exists today.)

The potential mass of all energimes (everything that emits gravitation) is 3.44×10^{58} kg. If you look at the ratio of the total hydrogen mass to the mass of the entire universe, you get 4.9 percent. In other words, scale metrics predicts that atoms make up only around 5 percent of all the stuff that emits gravitation! The remaining 95 percent is in the form of bound energimes that have not coupled with space to form matter.

Next up is an accelerating rate of expansion.

46

IS THE RATE OF EXPANSION ACCELERATING?

This topic is somewhat difficult because of the fundamental differences between scale metrics and the standard model. After a period of struggle, it became clear that I needed to stop trying to fit a square peg into a round hole. Within scale metrics, the universe is defined by one's square of the universe. As such, there is no stretching of the universe, and therefore trying to explain the accelerated rate of expansion is a task that has no meaning!

This is an important lesson for anyone interested in developing new ideas and new models. Often times, theorists become somewhat intimated by observation. They attempt to design their models (either intentionally or subconsciously) so that they agree with observation. Models that agree with observation are viewed as better and are valued as being more relevant. One needs to be careful about this tendency. For example, what would have happened if Einstein resisted the urge to introduce the cosmological constant? Remember, this fiddling with his model was done solely so that the results would agree with the prevailing thought of his day—namely, that the universe was static and not getting any bigger or smaller with time. If Einstein had held to the initial—simplest—version of his work, he would have most likely come to the conclusion that the universe was indeed getting larger. This may not have been received well initially, but with time—upon the observations of Hubble and others—it would have been viewed as a powerful prediction of his theory that was verified through observation.

This presents a kind of tug-of-war between theory and observation. It is not always obvious which camp has the upper hand. Observations are improved with time, and sometimes flat out errors or misinterpretations lead to modifications in what we observe (or what we think we have observed). Theories are also subject to change as new ideas are brought forward, misconceptions are corrected, and ongoing refinements are made.

So when it comes to scale metrics, there is absolutely no point in attempting to see whether it can predict the accelerating rate in the recession velocity of distant galaxies. This simply is not part of the scale metrics model. And, as shown in chapter 42, scale metrics can explain the observed redshift without requiring galaxies to be moving away from each other. Rather, light emitted in the past was governed by a different metric than the one we experience in the present.

Yet even though scale metrics presents a different model, it is still worth attempting to explain why the standard model comes to the conclusion of an accelerated rate of expansion.

To fully comprehend this, we need to better understand what is meant—particularly in modern cosmology—by the term "observation." What is observed is not necessarily what has actually happened. We have already discussed this in terms of proper and coordinate scales and the ongoing changes to the proper metric that—as an observer—you are completely unaware of.

But it goes further than this. One has to be careful with what one means by the word "observation." For example, a team of astronomers might state that they have observed a redshift in the light received from a far-off galaxy. Have they? The observation they made is that they saw light from a distant galaxy with a particular spectral signature. That is the observation. The claim that they observed a redshift is only true when they interpret their observation within the framework of the standard model. I like to call these model-based observations. They are not observations at all. They are observations that are influenced by a particular model of the universe. I stated this once earlier in the book, but I will repeat it. In cosmology today we have come dangerously close to treating GR and the standard model as if they are absolute. Therefore, it has become generally accepted to simply treat any model-based observation, when viewed within the framework of GR or the standard model, as if it were an actual observation.

This type of thinking can be very dangerous within science. We must be diligent in separating the observation from the model. Of course at any time you can introduce a model, but my point is, do not call a model-based observation an observation, because it is not one.

When you look more deeply into this, you quickly find out that the whole topic of cosmic expansion is a bit more complex then you originally might have thought. We start with Hubble observing that objects far away are seen to have light that is shifted to the red side of the spectrum. The farther away the objects are, the greater the redshift. This seems to be due to a Doppler redshift caused by galaxies moving away from us. But within the standard model, the galaxies are not really moving; rather, it is the fabric of space and space-time that is stretching because of a change in the properties and geometry of the GR metric. This means that the relationships that define the Doppler effect are not appropriate to apply to cosmic redshift. Yet many books and scientists continue to this day to utilize Doppler redshift in

FROM FALLING APPLES TO THE UNIVERSE

their analysis to explain the recession velocity of galaxies. This is both confusing and inaccurate.

Distant galaxies are not moving away from us in any real sense. Think again about the raisin bread in the oven. The Doppler effect would be the correct way to look at redshift if the raisins where actually moving through the dough. But in the standard model, the raisins do not move relative to their positions in the dough (space); rather, the dough simply stretches outward. This means we cannot simply apply the Doppler equation, even in its relativistic form, and expect to get accurate results.

But if we cannot use the Doppler equations, how then can we evaluate and make meaning out of redshift observations? You must use a model. And that means your assessment is only as good as the model you are utilizing. It also means, by definition, that all your observations are indeed model-based observations.

So what do scientists think they know? The Hubble value can be used to predict the age of the universe. This is a universe that is—more or less—coasting along with a constant recession speed. This is the case where a doubling of time will always result in a doubling in the scale of the universe. A note for the more advanced readers: This is not the same as to say that the Hubble value remains constant. The inverse of the Hubble value provides the age of the universe. So, indeed, the Hubble value actually changes with time yet still predicts a uniform coasting of cosmic expansion with no influences that cause any acceleration or deceleration.

When you apply GR to this, you get a slightly different result. In GR, all the mass and energy of the universe should be applying a gravitational influence that with time slows the rate of recession. Therefore, you would expect to see distant galaxies as having been receding faster in the past, resulting in distant light appearing to be brighter than expected simply by the Hubble value.

The observations from 1998 by two independent teams measuring the brightness of light emitted by distant supernovae suggested something still different. They actually measured distant light—from about five billion years ago—to be about 25 percent dimmer than expected. The model-based interpretation of this observation is that it is due to an accelerated rate in the recession velocity of galaxies. This model-based observation is truly amazing because—based on GR alone—it is the opposite of what should have been expected. That is, distant light appears to us to be dimmer, not brighter, than expected.

Think of it this way: You are on a road trip and plan on reaching your destination in two hours traveling at a constant speed of 60 mph. This is like the Hubble prediction—nice and steady. Yet what actually happens is that you run into road construction for the first hour, which slows you way down. Eventually you get back to your target speed of 60 mph, and then you speed up even more during the second hour of the trip and still reach your destination in the two hours you had planned. You did not really go 60 mph throughout the trip, but it still averaged out that way. But what are the odds that your trip would work out perfectly like that? If your destination had been closer or farther away, it would not have worked out so nicely. Yet this is how the standard model works. There is a slowing down of the recession velocity of the universe followed by a period of coasting and then—about five billion year ago—an acceleration in the rate of recession. And when all averaged out, you get the Hubble value. Once again, we just happen to live at that very special moment in the history of the universe when these values conveniently average out the right way!

And this is all achieved by applying model-based observations using GR and the standard model. However, in reality, observing the spectral signature of far-off galaxies does not conclusively tell us any of this. All we really know is that observations were made of the spectral wavelengths from light emitted in the past.

Scale metrics provides a much different model-based explanation. The light that you see today from galaxies far away was emitted in the past. These photons encountered all the free energimes defining the distance between the source and you, the observer. As these photons traveled through distance and time, the metric of the universe also changed. As a result, the photons traveled through all of the energimes defining the distance at the time the light was emitted, plus half of all the additional free energimes emitted along this path during the time that transpired. In other words, the path traveled by the photons was partly defined by the metric at the time the light was emitted and partly by the current metric of the observer. Yet the energy per photon was set—fixed—based on the metric at the time the light was emitted. And the observed distance traveled is determined solely by the current value of the metric. In scale metrics, adding more space does not change the value of the photon's energy, but it will change the value for distance. Because of this, the distance observed by a current observer is greater than the actual distance traveled over time by the photons.

When using scale metrics, light seen with a z value in the range of 0.5–0.6 will be observed to be about 11–12 percent farther away than anticipated. Since the intensity of light changes with the square root of the distance, this suggests that the light will be seen to be about 23–25 percent dimmer. This is exactly what was observed by the two independent teams in 1998! Yet it does not involve any stretching of space or space-time. We simply add more free energimes

IS THE RATE OF EXPANSION ACCELERATING?

with time, thereby changing the value of the metric and increasing the distance an observer will see without changing the energy of the light emitted.

The point is that the same observation can support more than one model. It is only when we lock ourselves into thinking that a model-based observation is a true observation that we become limited to a particular way of viewing that observation.

In any case, scale metrics provides an explanation for what appears to be a changing rate of recession without requiring any recession at all! This leaves open the question of whether space is indeed stretching or not.

47

IS SPACE STRETCHING OR NOT?

One of the biggest contrasts between the standard model and scale metrics is that one says the universe came into full existence within an instant while the other says mass, distance, and time (as space) are gradually added into the universe. And as you know by now, this space comes from the phase change of bound energimes to free energimes.

Before we move on, let's take a break and make some popcorn.

Now in the standard model of popcorn making, there is one kernel of corn that at some point—pops! This is all the popcorn that you get; it is the entire universe. And as time goes by, this piece of popcorn, which came into being in an instant, will stretch and eventually define all the space in the universe as we see it today.

When we make popcorn using scale metrics, there are many kernels of corn, each representing a singularity of its own. These are the bound energimes of scale metrics. As time goes by, more and more kernels pop (phase change), increasing the volume of popped corn in your bowl. The universe does not "pop" into full existence in an instant and then simply stretch. Rather, the components of the universe—mass, distance, and time—are released in a gradual manner, resulting in an ever-increasing amount of space

FROM FALLING APPLES TO THE UNIVERSE

within the universe, which currently defines the universe as we see it today.

Why is this distinction important? For one reason, if we use the stretching-of-space concept, we must ask what the universe is stretching relative to. This is a fair question to ask. When given the raisin-bread analogy, you often get the following narrative: You are told to watch the dough rise and the raisins separate (this observation requires that you are outside the loaf of bread). You are then told to imagine that you are inside the bread, and you will then experience the concept of the standard model.

The problem—as I see it—is that as you put yourself inside the raisin bread, you are carrying with you into the bread some notion of what distance is. That is, if you had a ruler with you on the outside to measure the dough expanding, you inadvertently carry this ruler with you into to the bread. This is now the standard from which you measure the stretching to occur. This means that even from within the bread, you are aware of some special dimension that supersedes whatever is actually occurring within the bread. After all, if all your experiences were only from within the bread, then your ruler would stretch along with the dough and you would have no knowledge of any stretching. You would simply be experiencing a uniform change in scale. Your metric would be changing, but you would be totally unaware of it!

Seems like a problem, but to the credit of the standard model, there is a response. The response is that your ruler will not stretch, because matter is not subject to the stretching of the metric in the same way that space is. OK, let's play along. Matter is different than space, and only space is stretched; matter is not. So my ruler can measure the stretching of space because my ruler is universally constant and unaffected by any stretching of space that occurs.

This extends not only to atoms but also holds at the level of galaxies. That is, galaxies do not stretch either with the stretching of space. They stay essentially unchanged as the space between galaxies stretches.

But isn't most of a galaxy made up of space? To demonstrate this, most models that simulate a collision between two galaxies predict very little actual impact because most of the matter in each galaxy will simply pass right through the other galaxy. Why? Because galaxies are mostly empty space with matter sprinkled within. Why wouldn't this space stretch just like the rest of space outside these galaxies?

Well, there is no real good answer to this. Or, as scientists like to say, this is a difficult and challenging topic. The general idea is that gravitation—over short distances relative to the size of the entire universe—is able to counter the effects of stretching space. This is a bit artificial in that gravitation, as defined by either GR or scale metrics, is invariant to time. The value of the metric has nothing to do with the passage of time and therefore is independent of any stretching that would occur over time. And even if you attempted to factor this into gravitation, it would suggest the existence of a "sweet spot" where gravitation would exactly counter the effects of stretching. A balance point would exist where gravitation and dark energy cancel each other out. At distances less than the sweet spot, gravitation would dominate. At distances beyond the sweet spot, the effects of dark energy would dominate. While a galaxy is much smaller than the universe, one would still expect to see the differences within this tug-of-war between gravitation and dark energy within a galaxy when played out over billions and billions of years. Yet scientists see no such effect.

This immunity to stretching is also seen within atoms and molecules. They do not change in size with the stretching of space. Therefore, the objects made with atoms and

FROM FALLING APPLES TO THE UNIVERSE

molecules—like meter sticks—also do not stretch with time. But, once again, most of what makes up an atom is empty space. Take the hydrogen atom, the most prevalent atom in the universe. It is made up of a proton with an electron existing in a large cloud outside the nucleus. Most of the hydrogen atom is indeed empty space. So apparently the electrostatic influence also perfectly balances dark energy. Yet the electrostatic influence is 10^{42} times stronger than gravity. If gravity is set just right to counter stretching within the scale of a galaxy, how can a different influence 10^{42} times stronger also exactly balance dark energy at a much smaller scale?

How does scale metrics address this? Within scale metrics, space is added to the universe with time as bound energimes phase change to free energimes. This occurs everywhere—all the way from the vast expanse of the universe down to the tiny scale of the atom.

But within scale metrics, one must always be aware of the metric being used to make the measurements. My metric today is much different than the metric that governed the universe billions of years ago when the components of a galaxy were indeed much closer together. And yet if we allow things that were close together in the past to move apart with time but then measure their distance using only the current metric, we observe the distance across the galaxy of the past just as we would observe that galaxy today. This not only holds for galaxies but also for atoms and molecules, and it has absolutely nothing to do with gravitation or electrostatic influences.

Look at it this way: Nothing changes on the coordinate placement of matter within your square of the universe. Two objects remain, for all time, at the same location unless influenced by gravitation (or electrostatics). And if we used

IS SPACE STRETCHING OR NOT?

our current metric billions of years ago, we would see them to be the same distance apart as they are in the present.

The bottom line is, space does not stretch; we simply add more space into the universe with time.

48

A NOT-SO-HOT BEGINNING

Last on the agenda is a discussion on the temperature of the universe. This again is an area where scale metrics would appear to be way out of line. A model that suggests a constant temperature to the universe for all time must certainly be wrong. Why? Because supposedly we have proof of a hot, dense beginning to the universe as observed by the CMBR. But do we?

Once again, any "proof" of the big bang based on the CMBR is couched within the subtlety of a model-based observation. All we know through observation is that uniform microwave radiation can be detected from any direction within the universe. That in and of itself tells us absolutely nothing about how far that radiation has traveled, when it was emitted, or the temperature of the universe at the time it was emitted. The idea that the CMBR suggests a hot, dense beginning to the universe only applies when the observation is influenced by the standard model.

To make this point, consider the following: Your house, on a cloudy day with steady outside temperatures, will reach an equilibrium temperature inside your home. And when this occurs, the walls of your home will emit thermal radiation consistent with the temperature of the air molecules within your house. At equilibrium, each of the walls from any direction within your home will emit the same wavelength of radiation. It does not matter how far the radiation traveled, nor does it matter which direction you look. All the walls will emit the same radiation (assuming we can treat your home as a black body, meaning the walls of your home are emitting all the energy they are absorbing).

That is the observation you will make. Now I ask you, from that observation, do you immediately conclude that your house was smaller in the past and that the walls of your home are stretching outward? Do you immediately conclude that in the past, the temperature of your house was much hotter and has only now cooled to its present temperature through an ongoing adiabatic expansion? Do you conclude that your home burst into existence long ago from an initial singularity? Absurd conclusions, right? You would only come to these conclusions if you were interpreting your observation as influenced by some model, such as the standard model.

And, according to this model, in the beginning the universe was way too hot for atoms. It needed to cool to the point that electrons and protons could come together primarily in the form of hydrogen and helium. Further, the universe needed to cool to the point where it became transparent and light could actually be emitted that would travel across the universe (as opposed to being almost immediately reabsorbed). The surface of last scattering is the theorized source of the energetic CMBR that we see today in the microwave spectrum after nearly fourteen billion years of cooling caused by the stretching of space and space-time. Once again, as is often the case in modern cosmology, this all happened within an instant (at least relative to cosmological time) when the universe was 380,000 years old. The radiation has supposedly been traveling outward into the universe for nearly fourteen billion years to what we see today. So according to the standard model, when we detect the CMBR we are actually seeing nearly fourteen billion years into the past. How do we do this? We look to areas of the universe where nothing is present. That is, if we look directly at the sun, a star, or a galaxy, we cannot accurately measure the CMBR. It is only when we look into the vast expanse of nothing all the way back to nearly the beginning of time that we see the CMBR.

Scale metrics also tells a story of the initial history of the universe. It is a much different story, but it is firmly based within the model of scale metrics. When we view the CMBR within the framework of scale metrics, we end up with a much different model-based observation and we draw much different conclusions.

We know from scale metrics that all matter comes from a coupling of free and bound energimes and that the quantity of space in the universe is what determines the total amount of matter that will be present. Of course, at the beginning there was no space and therefore no matter. There are only the singularities represented by 1.58×10^{131} bound energimes that have not yet phase changed to free energimes. We also know that not all mass was introduced in an instant but rather that matter continues to form within the universe as more free energimes become available. And all these interactions occur at a constant and equilibrium temperature of 2.7K.

To be more specific, the first electrons appeared when the electron's Compton wavelength squared divided by the inverse fine-structure number resulted in an electron mass that could be supported by the early universe. Remember, in scale metrics the mass of the electron and the inverse fine-structure number both change as the universe evolves. Not everything came into existence in an instant. The universe was not born with one big bang, and the electron did not appear in a "poof" and then remain exactly the same for all of history. Using scale metrics, the mass (and therefore charge) of the electron continues to evolve with the evolution of the universe.

So in order to have a stable electron, one whose existence can be supported by the early universe, you need to have enough free energimes to support the mass of the electron. For that electron to be real to us, it must also be stable within the portion of the universe that we can directly interact with.

And when would that occur? Based on the criteria above, the first electron had a mass in the area of 10^{-55} kg. This is roughly 10^{25} times smaller than the electron of today. This electron formed very early when the universe was around 10^{-3} to 10^{-4} of a second old and contained an energy consistent with a thermal temperature of 2.7K. Since this time, more free energimes are present, allowing for increased amounts of matter to form within the universe. And as additional matter forms throughout the ages, it also emits radiation at a thermal signature of 2.7K. In scale metrics there was no surface of last scattering that occurred at 380,000 years and that generated all the CMBR. The CMBR is emitted throughout all ages and has traveled various distances to reach us today. But regardless of the direction it comes from or the distance it has traveled, it bears a signature temperature of 2.7K that is consistent with the temperature of our universe across space and time.

Temperature is a measure of average kinetic energy. The radiation given off as a thermal signature has nothing to do with the electron moving between energy levels within an atom. It has to do with the vibration of a charged particle brought about by the kinetic energy of the particle. Scale metrics states that that average kinetic energy has always been at 2.7K. The standard model states that the early universe was intensely hot, with temperatures around 5,000K at the time the CMBR was emitted. How can two such drastically different ideas both lead to the universe we see today?

It takes work to build the universe. The word "work" in science is not that much different from how we think about it in daily life. Work can be thought of as the total amount of energy necessary to complete a task. There are two extremes in how you might complete a task. You can exert a lot of effort over a short time. Or you can work at a more relaxed pace over a longer period of time. Why is that important to mention? Because based on scale metrics, what we experience as one second today was actually 10^{61} seconds in the first instants of the universe. A lot of work can get done, even at a low energy level, when it is extended over a long period of time.

Think of it this way: Let's say the holidays are going to be celebrated at your house this year. If you know well in advance that you are hosting, then you have plenty of time to get ready. You can proceed in a relaxed, low-energy manner. On the other hand, if Aunt Martha cancels at the last minute and you get the holiday by default, well, that is something quite different. When you find out at the last minute that everyone will be descending on your house, you run around like a crazy person trying to get ready. The result is the same, it is just a question of the path you take to get there: fast and furious or slow and relaxed.

Last example: Let's say you are watching a movie that features a building undergoing a massive explosion. If you watch that same footage in slow motion, you see something quite different. You see pieces moving outward at a very slow rate. The result is the same, but by watching it in very slow motion, you would hardly call the event an explosion. It appears as a much lower-energy event that plays out over a much longer period of time.

In scale metrics the universe has been around for a long, long time from the coordinate scale (roughly our proper time squared). It did not have to be in any hurry to evolve.

It had plenty of time to reach an equilibrium, to form matter that continues to come into existence and evolve with time, and to ultimately arrive in the manner we observe today. There is no need for an intensely hot beginning.

So in the end, scale metrics suggests we live in a universe that has always been cold and that is fueled by the ongoing phase change of energimes from bound to free states within an isothermal process—a simple yet very effective model for simulating and mimicking what we see (as long as we remain open to how we interpret our observations).

49

CONCLUSION

For many of you, this may be the end of our journey together. Thank you for taking the time to participate. Each of you will take away what you want from this experience; but, if I may, I would like to suggest the following highlights.

If nothing else, walk away from this book with the notion that space may be much different than what we ever thought. Space is not just the void between two physical objects. Space is—in a very real sense—everything. All the mass of the universe comes from space. All the distance of the universe comes from space. And all the time of the universe comes from space.

We live in a universe that may not have started with a big bang but rather with space being introduced gradually over the full history of its evolution. As such, the universe is not expanding; there are simply more free energimes phase changing into existence and defining an increased quantity of space (mass, distance, and time) while maintaining a constant temperature.

Scale metrics introduces a universe that can be explained within our three dimensions of physical awareness and within the framework of Euclidean geometry. A universe that can be understood by everyone through the use of language and with mathematics used to support the narrative (section 3).

FROM FALLING APPLES TO THE UNIVERSE

A universe whose overwhelming size is required so that we can experience things just the way we do here on Earth. A universe where gravity and gravitation are recognized as being different and where motion is seen as a static property of matter yet always requiring energy (even across an infinitesimal interval).

A universe where nature has beautifully set limits requiring an infinite amount of energy to achieve the speed of light and an infinite amount of mass to achieve an event horizon. As a result, gravitation does not support the current notion of black holes.

A universe where the concept of unity is defined by nature itself, not by humankind. A universe that reduces to a fundamental particle that which defines the unity values of mass, distance, and time as manifested through space. As such, all measurements and physical constants can be reduced to the simple counting of energimes (while keeping track of what you are counting: mass, distance, or time).

A universe that rejects the notion that all observers (within a small region of space and space-time) have a valid claim on being the observer at rest or that all physical events can be accurately described from one observer's at-rest frame of reference. Rather, all motion is relative to space, the ultimate coordinate system (grid) from which all events play out.

The above are all features of scale metrics, a model—using the simplest form—that mimics, simulates, and explains many of the observations we see in the universe and that provides answers to a number of puzzling questions. Scale metrics predicts a universe

- with an age of roughly fourteen billion years;
- that is homogeneous and isotropic without the need for a special inflationary epoch;

CONCLUSION

- in which only 5 percent of all the stuff responsible for gravitation is in the form of atoms;
- that does not stretch with time but rather may appear to be accelerating only because of our insistence in applying our current metric to all things we "observe";
- where the inverse of the Hubble value always provides the age of the universe (we do not live in some "special moment");
- that explains the strange value of the inverse-fine structure number (since there is nothing special about "right now," there is no reason to believe that the current value of the inverse fine-structure number has any special cosmic significance); and
- that is composed of energimes, with each energime containing the unity values of mass, distance, and time.

Scale metrics further suggests a universe that is undergoing constant and gradual change as many of its features evolve with time, including

- the mass of the electron, proton, and neutron;
- the value of the inverse-fine structure number;
- the value of the metric;
- the speed of light;
- the atomic spectrum; and
- the quantity of space (mass, distance, and time).

Is scale metrics correct? That is the wrong question to ask! Scale metrics is nothing more than an idea that was developed into a model. Is it a good model? Well, if a model is tested against how well it mimics and simulates what we observe, then yes, it is a good model. If a model is tested against whether it can make predictions that agree with observation without constant fixes, patches, or adjustments, then it is an even better model.

Yet while the above features demonstrate some of the achievements of scale metrics, scale metrics is far from a description of reality. Always remember that while science

FROM FALLING APPLES TO THE UNIVERSE

is the search for reality, it seeks a goal it will never achieve! That is partly the beauty of science. The journey never ends.

There will be many new journeys in the future. It is my hope that scale metrics might serve as a starting point for one of them. Perhaps a journey that you will undertake as humanity continues its search to understand the universe.

SECTION 3
UNDERSTANDING
THE MATHEMATICS

50

INTRODUCTION

Welcome to section 3. We will be taking another signifi-cant turn, as I now look to provide a more mathemat-ical story line that supports and justifies the narrative from sections 1 and 2. Collectively, all the referrals to section 3 made in sections 1 and 2 will be addressed. However, it is not my intent for section 3 to simply be a resource; rather, my intent is to provide a comprehensive (and—yes—more mathematical) description of scale metrics.

While scale metrics may appear to be a significantly different model of gravity and gravitation with far differ-ent predictions on the nature of the universe, it is really grounded in only a few quantitative concepts. Of course the most fundamental of these is the notion of the energime that exists as a segment in free space (once phase changed) and that defines the unity values of mass, distance, and time. We have collectively come to know the combination of these properties as space.

Next, there are some specific questions to answer: Why is it that scale metrics treats h as the heat-death radius of the universe? What is the true relationship between G and h? How does scale metrics determine the mass of the space frame to be $2\pi hr$? How does one determine the scaling fac-tor used to calculate the velocity of an object? What was the composition of the electron (and other matter) in ages long ago? How does scale metrics quantitatively address the observation of an accelerating expansion rate to the universe?

Beyond questions like these, the single most important aspect of this book—beyond its robust treatment of space—is its derivation of a general form of the kinetic energy equation. While the kinetic energy equation was introduced along our journey into scale metrics, it is important to note that it actually stands on its own merit and does not depend on the success of the scale metrics model. Therefore, even if you find some aspects of scale metrics to be speculative, you cannot dismiss the implication that the mathematical equations of GR may require modification. Interestingly, once these modifications are made—entirely separate from scale metrics—the GR equations become fully aligned with those of scale metrics. Perhaps this provides some of the strongest evidence in favor of the scale metrics model.

Finally, one cannot lightly reject the notion of general covariance. Therefore we will be looking more deeply into this along with both special and general relativity. We will review these ideas against the backdrop of physical observation.

In the end, scale metrics provides a powerful new view of the universe, one that has been explained through the use of both language and mathematics. And, just as with GR or any other theory, scale metrics will be nothing more than a model that mimics and simulates physical observation. Like all of science, scale metrics will be part of the search for the reality it will never find.

Let's begin by looking into the notion that h represents the heat-death radius of the universe.

51

PLANCK'S CONSTANT AS A DISTANCE

Let's start with something fairly easy, straightforward, and somewhat fun.

Planck's constant is generally viewed as a quantum entity that relates the energy of a photon to its frequency. Scale metrics has made the claim that h can also be viewed as the heat-death radius of the universe. Or, by applying the speed of light as unity, h can also be used in determining the age of the universe (chapter 43). While these may seem like vastly different statements, it is actually quite easy to derive h as a distance (or a time). We know that in scale metrics units can be somewhat misleading and that all constants can be reduced to nothing more than a dimensionless number. The dimensionless value of h is 9.95×10^{64}, which can also be expressed in units of distance as 1.01×10^{31} m. This is accomplished by multiplying the dimensionless value of h by the unity value for distance, which is 1.02×10^{-34} m. However, h can equally be expressed in any combination of the energime values of mass, distance, and time. So the basic question becomes, What is the most fundamental way to express Planck's constant?

We know that

$$E = h\nu = \frac{hc}{\lambda},$$

where E is total energy content, h is Planck's constant, ν is frequency, λ is wavelength, and c is the speed of light. We also know that energy is typically expressed in units of joules, and when this is done, h must be in units of joule-seconds for the units to be consistent on both sides of the above equation. Yet when using scale metrics, we know that total energy content is simply the sum of energimes present. The energy of a photon is simply the number of energimes within the photon. We also know that wavelength, by definition, is a distance and that the speed of light, by definition, is unity. Therefore, when using wavelength as the fundamental property of a wave, h must be expressed as a distance when we express energy as a dimensionless number.

In this sense, viewing h as a distance is every bit as valid as any other interpretation. In scale metrics we will define the most fundamental way to determine energy as the counting of energimes and the most fundamental way to define a wave as by its wavelength. Based on these definitions, h as a distance is the most fundamental way to express Planck's constant.

Next, once we realize that the universe can be no bigger than the Compton wavelength of the smallest quantum, it follows that not only can h be viewed as a distance but also as the largest possible distance that can be observed within your square of the universe. You might suggest that since there are a large number of energimes in the universe, you can simply string them together to make larger distances. However, you must realize that all events occur within your square of the universe. What I am suggesting is that at the moment the last bound energime phase changes to space,

PLANCK'S CONSTANT AS A DISTANCE

the observed distance across the universe will be h, and this will represent the wavelength of the last energime to undergo phase change as viewed from the proper scale.

Of course, we are not at heat death yet. In fact, according to scale metrics, we are far from it. How then can we treat h as a constant when it represents something that has not yet occurred? The answer is that the distance across your square of the universe is—currently—only a portion of h and that this same proportional value applies to the measured wavelength of any photon in today's universe. If both h and the wavelength of a photon change proportionally to each other, we may use h as a universal constant for all ages. This is consistent with the general notion in scale metrics that most things are changing with time.

If you want to have some additional fun with dimensions, h can also be viewed from the perspective of frequency. In chapter 43, we derived the frequency of the energime as 2.97×10^{-23} cycles/s, which is equivalent to the dimensionless number 1.005×10^{-65}.

So we see that the value $h\nu$ for a single energime also returns the value of one, which is the energy of a single energime:

$$E = \left(9.95 \times 10^{64}\right)\left(1.005 \times 10^{-65}\right) = 1.$$

In keeping true to expressing h as a distance, we can also express the energime frequency as 9.90×10^{-32} cycles/m.

This may look awkward or even incorrect, but within scale metrics we can express a value in any units we choose (or no units at all). We simply need to keep track of what we are measuring. So, yes, we can express a frequency in units of cycles per meter if we wish.

And once again the value of $h\nu$ is determined to be one.

$$E = \left(1.01 \times 10^{31}\,\text{m}\right)\left(9.90 \times 10^{-32}\,\text{cycles}/\,\text{m}\right) = 1.$$

We see that h, as well as any other constant or measured quantity, can be expressed in various combinations of mass, distance, and time. Yet if we are willing to define a quantity of energy by the number of energimes and a wave by its wavelength, we must then conclude that h is best defined as a distance. And, at the end of days, it will indeed be the heat-death radius of the universe as observed from the proper scale.

Knowing this, let's now turn to the relationship between G and h.

52

THE RELATIONSHIP BETWEEN G AND PLANCK'S CONSTANT

In chapter 12, both the gravitational constant G and Planck's constant h were derived from the unity values of the energime. This was based on the assumption that G and h are inversely proportional to each other, a statement that needs to be quantitatively confirmed.

I also want to look closely at the physical significance of both G and h and in so doing establish a clear conceptual and quantitative relationship between the two values. This goes beyond claiming they are universal constants but rather goes to the heart of what they actually represent in a physical and tangible way. In the process, we will also establish the scale metrics equation for the value of the metric at any distance r from a gravitating mass as well as the mass of the space frame as $2\pi hr$.

Further, I want to touch on the topic of universal constants. Much of this book has been focused on the notion that most things change with time. As suggested in the last chapter, h continually changes with time in a proportional manner so that it simulates a universal constant. If G and h are indeed related, it follows that the same is true for G. The gravitational constant is not truly a universal constant but changes proportionally as space is added within the

FROM FALLING APPLES TO THE UNIVERSE

universe. However, just as was the case for h, space changes in a manner proportion to G such that G may also be treated as a universal constant.

In scale metrics the value of the metric is literally the space (whether looking at mass, distance, time, or all three combined) between adjacent free energimes. A gravitating mass emits free energimes outward from its center into the overall large-scale structure of the universe, thus altering the value of the local metric. The contribution of a gravitating mass to the overall metric is determined by only two variables: the mass of the gravitating body and the mass of that portion of the universe involved in the interaction.

In other words, I can view the large-scale structure of the universe as a large grid defining my square of the universe. If I wish to determine the value of the metric b at a distance of r, I only need to account for the portion of space necessary to reach a distance of r from a gravitating mass. This idea was introduced in chapter 19 as the space frame. It is the same value that is used in conjunction with the scale metrics equivalence principle.

Therefore, the metric takes on the form

$$b = \frac{M_{(\text{space frame})}}{M_{(\text{space frame})} + M_{(\text{gravitating body})}}.$$

Here is where scale metrics reinforces the concept of the energime segment. While I can consider only a portion of the grid within my square of the universe, I cannot break up the energime segment defining the third dimension. I ask you to once again return to the trees in the forest. I can certainly look at a smaller footprint on the forest floor, but if I wish to consider the volume defined by that footprint, I must take into account the full height of the trees. If we

THE RELATIONSHIP BETWEEN G AND PLANCK'S CONSTANT

view the segment as denoting time, then we conclude that while it is possible to look at only a portion of the cross section of the universe, we must include all the time in which the universe has existed. We cannot pretend the universe is a different age than it actually is. There is no degree of freedom to move up or down the time dimension; we must take time exactly as it exists in the present (chapter 28).

Any portion of the universe along the two-dimensional grid can be defined by πr^2, where r defines the radius of a circle defined by the distance between the gravitating body and the object under the influence of gravitation. However, the actual radius of the universe (and thus the entire segment of time) is defined as $2h$ (chapter 32). This means that the total space involved in an interaction at a distance of r is stacked along the energime segment by the ratio of

$$\frac{2h}{r}.$$

Therefore, the total portion of the universe considered in determining the metric at a distance of r is

$$\left(\pi r^2\right)\left(\frac{2h}{r}\right) = 2\pi hr.$$

The metric b then takes on the form

$$b = \frac{2\pi hr}{2\pi hr + M_{gravitating}}.$$

Dividing through by $2\pi hr$ gives

$$b = \cfrac{1}{1+\cfrac{M}{2\pi hr}},$$

where M is the mass of the gravitating body and r is the distance of separation between the gravitating mass and an object under the influence of gravitation.

Further, in defining the gravitational constant as

$$G = \frac{1}{2\pi h},$$

we achieve the scale metrics definition of b as

$$b = \cfrac{1}{1+\cfrac{GM}{r}},$$

and we also confirm the inverse relationship between G and h. These two constants take on very tangible meanings. As first suggested in chapter 11, if h is viewed as the radius of the universe, then $2\pi h$ represents its circumference and G is the inverse of the circumference of the universe.

We now better understand the value of the metric, the space frame, and the relationship between Planck's constant and the gravitational constant. Let us now turn our attention to momentum and the process for determining the scaling factor.

53

MOMENTUM: DERIVING THE SCALING FACTOR

I n chapter 17, I suggested that the velocity of an object can be determined from the momentum transferred during a collision along with a scaling factor that reconciles the motion of the energime defined by the coordinate scale with the speed of light as defined by the local scale. There are two aspects to this: the difference in scales and the difference in orientations, x.

At the first moment of the universe, the coordinate and proper scales were identical, both defined by your square of the universe. As the universe evolved, free energimes were emitted, which changed the ratio of the coordinate scale to the proper scale. That is, your square of the universe began to look like a grid and that grid became more tightly woven as time went by. The result is an ongoing change in the ratio between the coordinate and proper scales.

A second contributing factor is the orientation, x, of the energimes within the local scale. For an object with a velocity approaching zero, the value of x approaches one half. At this orientation, an energime transferred in a collision is equally likely to be traveling vertically as it is horizontally. As such, the net effect is a reduction along the line of motion equal to $\sqrt{2}$. These two factors are shown in the following illustrations.

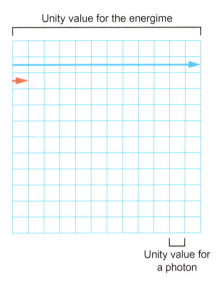

The energime motion vector (blue)
is defined by the global metric defining the universe.
The motion vector for light (red) is defined by the local metric.

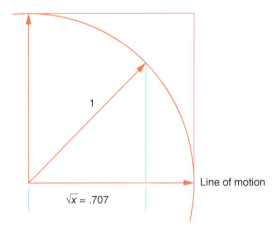

Lastly, we must be able to account for how an object obtains its kinetic energy. Yes, the receiving object experiences a momentum transfer due to a collision with an incoming object. But how does the initial object obtain its kinetic energy? To answer this, we can invoke the scale metrics equivalence principle. We can view an object's kinetic energy as being equivalent to that of an object falling through a gravitational influence. As such, the scaling factor is determined not from the beginning of time (the entire universe) but rather that portion of the universe involved in the interaction, which is $2\pi hr + M_g$.

This gives us a ratio of the mass of that portion of the universe involved over the mass of the gravitating body. Since mass plays out across the two-dimensional cross section of the grid, it follows that the change in the one-dimensional metric brought about by this mass ratio will be determined by

$$\left(\frac{2\pi hr + M_g}{M_g} \right)^{1/2}.$$

Furthermore, taking into account the change in orientation x provides the expression for the scaling factor:

$$\text{Scaling Factor} = \left(\frac{2\pi hr + M_g}{xM_g} \right)^{1/2}.$$

Realizing that from the scale metrics equivalence principle $m_e = m_g$ and $m_o = 2\pi hr$, the velocity of any object may be determined by a transfer of momentum multiplied by the scaling factor:

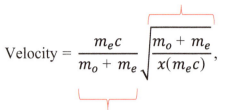

$$\text{Velocity} = \frac{m_e c}{m_o + m_e} \sqrt{\frac{m_o + m_e}{x(m_e c)}},$$

Scaling factor (predicted by scale metrics)

Velocity (as predicted by transferred momentum only)

where $c = 1$.

Next let's explore how the value of the metric plays an important role in how the hydrogen atom may have changed throughout the ages.

54

THE STRUCTURE OF MATTER OVER THE AGES

In chapter 42, I suggested that cosmic redshift may not be the result of stretching space and space-time but rather the result of matter from the past emitting light with a different spectral signature (one that was redshifted at the time of emission). This requires a more in-depth look at the structure of the atom and how this structure may change with time.

If we look at the hydrogen atom and apply the Bohr model, we can gain some insight into how the hydrogen atom may have existed in past ages. Again, the Bohr atom is not a perfect model, but it does perform well with regard to the hydrogen emission spectrum. The variables at play are the inverse fine-structure number, the mass and charge of the electron, and the radius and velocity of the electron in each orbital.

Just as a quick review: Scale metrics predicts a doubling in the scale of the universe with a doubling of age as experienced by an observer in real time. If the universe is roughly fourteen billion years old, at seven billion years it was half its current diameter. We also know that the square root of the inverse fine-structure number decreases with the age of the universe. Therefore, at seven billion years scale metrics suggests an inverse fine-structure number of 194.

The mass of the electron at seven billion years can be determined by (chapter 37):

$$e_{mass} = \frac{\left(e_{compton\ wavelength}\right)^2}{194} = 3.71 \times 10^{42} \text{ energimes} = 8.12 \times 10^{-31} \text{ kg}.$$

The electrostatic charge and the value of Coulomb's constant are directly related to the electron mass, resulting in the following values:

- Electron mass: 8.12×10^{-31} kg.
- Electron and proton charge: 3.71×10^{42} energimes.
- Coulomb's constant: $8.01 \times 10^{9}\ N \cdot m^2/C^2$.
- Ratio of all above values to their current value: 0.891.

Now, let's build a Bohr hydrogen atom from seven billion years ago. The electron orbitals remain quantized by the value of n.

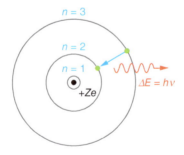

The size of the atom is larger than today's hydrogen atom, with the radius of each orbital given by

$$r_n = \frac{n^2 h^2}{ke^2\, 4\pi^2 m}.$$

THE STRUCTURE OF MATTER OVER THE AGES

If we normalize the value of the Bohr radius in our current age (fourteen billion years) to be one, we see that the radius of any orbital for the hydrogen atom seven billion years ago would increase by a factor of

$$r_n = \frac{1}{\left(0.891\right)^4} = 1.59.$$

The velocity of the electron also changes in each orbital:

$$V_n = \frac{h}{2\pi m r_1}\left(\frac{1}{n}\right).$$

And seven billion years ago it would decrease by a factor of

$$V_n = \frac{1}{\left(0.891\right)\left(1.59\right)} = \frac{1}{\sqrt{2}}.$$

Again, this is as compared with the hydrogen atom of today.

We immediately see that the Sommerfeld relationship between the velocity of the electron and the inverse fine-structure number holds perfectly true seven billion years ago. Just at the fine-structure number decreases by the square root of two, so does the electron velocity. This was by no means ensured to be true, so it becomes an important observation to make.

Next, we wish to compare the energy per energime seven billion years ago to that same value today. This step is taken to ensure that we are normalizing the comparison across a common metric. The photon properties will change with a changing metric. We want to eliminate this variable by focusing only on the constant value of the energime, whose properties are determined by the never-changing coordinate scale.

This suggests an energy per energime from seven billion years ago of

$$E_{energime} = k^2 e^4 \left(\frac{2\pi^2}{h^2} \right) \left(\frac{1}{n_f^2} - \frac{1}{n_i^2} \right),$$

or,

$$E_{energime} = \left(0.892 \right)^6 \left(\frac{2\pi^2}{h^2} \right) \left(\frac{1}{n_f^2} - \frac{1}{n_i^2} \right)$$

This suggests an amount of energy per energime between any two orbitals that is reduced by a factor of 0.5. That is, the energy per energime emitted seven billion years ago was exactly half the energy per energime that the hydrogen atom today emits across the same change in orbitals. We observe that today as a photon with a wavelength twice as long as the wavelength emitted by the current hydrogen atom.

This suggests it's possible that redshift is not due to the stretching of space and space-time but rather is accounted for by a past that was governed by a different metric. While this has only been modeled at seven billion years, it holds equally true for any age of the universe.

Yet this is not necessarily what Aged saw seven billion years ago. The point of scale metrics is that our universe is experiencing an ongoing change in the value of the metric. This change in the metric determines how matter existed in the past as well as how we see that past when it is viewed in the present. This differs from the standard model, which suggests that the structure of matter

THE STRUCTURE OF MATTER OVER THE AGES

has remained constant over the ages and that what we see today is the result of the stretching of space and space-time due to changes in the properties and geometry of the GR metric. Two very different approaches.

55

EXPLAINING THE EXPANSION RATE OF THE UNIVERSE

n chapter 46, I suggested that the scale metrics model is incompatible with the concept of an accelerating rate of expansion because there is no stretching of space or space-time within the scale metrics model. Some of this was addressed in the previous chapter, where redshift was explained through the notion of a different metric that governed the universe in the past. This metric allowed for light to be emitted with a spectrum that was already redshifted over what we would see today. But now we must ask why the standard model suggests an accelerating rate of expansion.

To address this with an example, place two objects within your square of the universe. Unless they are influenced by gravitation or electrostatics, they will remain there for all time. As space is emitted, each local observer sees his or her local frame to be changing, but the coordinate scale remains constant for all time.

What this means is that from the proper scale, the two objects are originally within a singularity (no distance between then at all). As bound energimes phase change to free energimes, the space between them changes as viewed from the proper scale but remains constant as viewed from the coordinate scale. As time goes by, the observer in real

FROM FALLING APPLES TO THE UNIVERSE

time sees the objects separate, yet the objects have not physically changed positions within your square of the universe.

This suggests a universe in which there is no stretching of space within a galaxy or within atoms. What we see today is the same as what we see (from our metric) from the past. But just as suggested at the end of the last chapter, what we see today is not necessarily what those in real time saw in the past. In the past, the objects were indeed very close together. As the metric has changed throughout the ages, we now see the past exactly how we would if the past existed in our current metric.

This means light that is reaching us today needed to travel through all the space in front of it plus half the new energimes emitted along the line of its journey. Its proper scale (as compared to the coordinate scale) was constantly changing over the entire journey. As shown in the previous chapter, it is this change in the value of the metric that accounts for what we interpret to be a redshift.

Now, let's look at light reaching us from four billion years ago. This is approximately the distance in the 1998 observations that led to the notion of an accelerated rate of expansion. Once again we are faced with the reality that redshift means nothing without a model from which to base our observation. Today this is generally done by applying the Friedmann-Lemaître-Robertson-Walker (FLRW) metric. To a first-order approximation, the FLRW metric suggests that the scale of the universe at the time a light pulse was emitted is

$$\frac{1}{1+z}.$$

Since the redshift in the 1998 observations had a value of $z = 0.5$, this suggests a universe that was roughly two-thirds its current scale at the time the light was emitted. If

EXPLAINING THE EXPANSION RATE OF THE UNIVERSE

we normalize our metric of today to a value of one, this suggests—based on scale metrics—that the metric at the time the light was emitted was

$$\sqrt{1+z} = 1.22.$$

Within scale metrics, the universe becomes larger (remember that there is no stretching) through two mechanisms (discussed in chapter 32): our awareness of additional free energimes (space) and the impact of these additional free energimes on the proper scale.

In other words, one meter today would have been measured as 1.22 meters four billion years ago if we were to apply our current metric across both measurements. Of course, the person experiencing the world in real time four billion years ago would have seen this distance to be one meter.

When viewed across a common metric—our metric of today—this changing value of the metric over time can be viewed as essentially linear. We can therefore take the average of the metric as it passed through the past four billion years to be

$$\frac{1+1.22}{2} = 1.11.$$

This suggests that we see the light as having traveled 11 percent farther than it actually did. Since the intensity of light changes with the square of the distance, this results in light that is roughly 23 percent dimmer, a value that is in good agreement with observation yet achieved without any stretching of space or space-time!

56

THE RELATIONSHIP BETWEEN KINETIC ENERGY AND VELOCITY SQUARED

Kinetic energy is the energy of motion. Classically it is defined as half the mass of an object multiplied by its velocity squared. More accurately, kinetic energy is the difference between the total energy content of an object possessing motion and its invariant mass. We rest comfortably realizing that the expansion of a Taylor series will show that these equations are in good agreement at speeds much less than the speed of light, c. This knowledge seems to have sufficiently addressed our needs for over a century. So why ask if we need anything more?

Somewhat ironically, our mathematical expressions for the energy of motion cannot be exactly expressed in a direct proportion to velocity squared. Why are we content with this? I suspect it is because we have become comfortable with a dual approach. For all of us, our first introduction was through classical mechanics, which works fine for normal-life situations. We later learned that at speeds closer to c, a relativistic adjustment was necessary. We have become complacent with the idea of using classical mechanics for speeds much less than c and including a relativistic

FROM FALLING APPLES TO THE UNIVERSE

adjustment for speeds roughly 10 percent that of light or greater. So why do we need to consider something new? Because our dual approach does not clearly and concisely define all the variables that equate kinetic energy to velocity squared. This chapter presents a clear derivation of an exact expression for kinetic energy in direct relationship to velocity squared.

At first glance, kinetic energy seems to be a straightforward concept. The idea that a moving object can perform work was experimentally tested in the 1700s by Willem 's Gravesande, who dropped metal balls onto soft clay and measured the displacement of clay from the impact. This relationship was expressed in a mathematical formula by Émilie du Châtelet that served as the precursor to what we express today as

$$E_{\text{kinetic}} = \frac{1}{2} mv^2. \tag{1}$$

For hundreds of years, equation (1) has been viewed as a natural law of nature and has become firmly footed within classical mechanics.

But in 1905, Albert Einstein's description of mass-energy equivalence as $E = mc^2$ suggested that energy itself exhibited the properties of mass. Therefore, kinetic energy could be more accurately expressed as the energy content of a moving object minus its invariant mass, or

$$E_{\text{kinetic}} = E - m, \tag{2}$$

where c is expressed in natural units with a value of unity, E represents the total energy content of the moving object, and m is the object's invariant mass.

THE RELATIONSHIP BETWEEN KINETIC ENERGY

The standard approach for reconciling these very different expressions is through a Taylor series. However, the value derived from a Taylor series is also an approximation by the very fact that only a limited number of terms are considered from an infinite series.

$$E = mc^2 + \frac{1}{2}mv^2 + \frac{3}{8}m\frac{v^4}{c^2} + \dots \qquad (3)$$

This creates an interesting situation. At speeds much less than c, it is generally accepted that kinetic energy is directly proportional to velocity squared. Yet while equation (1) includes a term for velocity squared, the equation is only an approximation of the actual kinetic energy as determined by equation (2). And while equation (2) provides an exact expression of kinetic energy, it lacks any clear proportional relationship between kinetic energy and velocity squared. Finally, the Taylor series from equation (3) includes various velocity terms but requires the summation of an infinite number of terms to determine the exact value of kinetic energy.

So is it possible to derive an exact expression for kinetic energy in direct relationship to velocity squared? Yes, as was suggested in chapter 15. We can go through its derivation right now. On the left side of equation (1), we can use the Einstein relationship for kinetic energy from equation (2):

$$E - m.$$

The right side of equation (1) is the classical equation for kinetic energy; however, it must also take into account relativistic effects and should include the terms γmv^2. Where γ represents the Lorentz factor,

$$\gamma = \frac{1}{\sqrt{1 - \dfrac{v^2}{c^2}}}.$$

Notice that the one-half term has not been included. I am replacing the one-half constant with a variable x that will make the expression an absolute equality.

$$E - m = xymv^2. \tag{4}$$

Now, if x can be clearly defined, then equation (4) will represent an exact equation for kinetic energy that includes a term for velocity squared.

We know that $E = \gamma m$, and therefore

$$\gamma = \frac{E}{m}. \tag{5}$$

We also know that if mass is to remain invariant to any frame of reference, it must be defined as

$$m = \sqrt{E^2 - p^2}, \tag{6}$$

where p represents momentum and c is defined in natural units to be unity. Momentum may also be expressed as total energy content, E, multiplied by velocity, v, so that equation (6) may be rewritten as

$$\left(Ev\right)^2 = E^2 - m^2.$$

And dividing both sides by E^2 yields

THE RELATIONSHIP BETWEEN KINETIC ENERGY

$$v^2 = 1 - \left(\frac{m}{E}\right)^2. \tag{7}$$

From equations (5) and (7), we can now define all the variables of kinetic energy in terms of E and m, as required by the Einstein relationship in equation (2).

Next, by substituting the defined variables into equation (4), x can be isolated and exactly expressed as

$$x = \frac{E - m}{\left(\frac{E}{m}\right)m\left[1 - \left(\frac{m}{E}\right)^2\right]} = \frac{E - m}{E\left(\frac{E^2 - m^2}{E^2}\right)} = \frac{E(E - m)}{(E - m)(E + m)} = \frac{E}{E + m}.$$

The variable x can be concisely defined as

$$x = \frac{E}{E + m} = \frac{\gamma}{\gamma + 1}. \tag{8}$$

Since

$$E = M_o + M_e,$$

where (M_o) is equal to the rest (invariant) mass and (M_e) is the mass of the additional energy added, it follows that

$$x = \frac{M_o + M_e}{2M_o + M_e},$$

as suggested in chapter 15.

Therefore, an exact expression for kinetic energy valid for all velocities can be written as

$$E_k = x\gamma mv^2 \qquad (9)$$

Notice that x approaches the classical value of one-half when E and m are near the same value, as shown by equation (8). This occurs at low speeds for which the energy content is very close to the invariant mass; or, when γ is very nearly one.

However, also note that x is a variable and that its exact value for an object in motion is never equal to a constant of one-half. This would require an object to gain velocity without increasing its overall energy content and would be a direct violation of mass-energy conservation. This confirms the earlier statement that kinetic energy as $1/2\, mv^2$ is never exactly correct. In fact, the classical equation is only correct in the very restrictive case when absolutely no motion is occurring. In all other situations, it is an approximation that we have allowed ourselves to use at speeds much less than c because it is so close to the exact value provided by equation (2).

Therefore x has values within the range of one-half to one but with open endpoints where one-half and one are not allowable values for any object in motion. That is, an object in motion can never be totally at rest nor will it ever achieve the speed of light.

Equation (9) eliminates the dual approach to kinetic energy as defined by both equations (1) and (2). It is an exact equation for kinetic energy at any velocity. We will find it to be a very powerful equation. First, it is a singular expression that is similar in form to the classical equation but is valid for any velocity. Second, it captures all the variables that determine kinetic energy. And third, based on the definitions of x, γ, and v, it clearly establishes that these variables are not independent of each other but that if one value changes, the other two values must also change. This interdependence of x, γ, and v has significant implications for gravity and gravitation.

57

IMPLICATIONS FOR GRAVITATIONAL REDSHIFT FROM AN EXACT KINETIC ENERGY EQUATION

For an object that gains kinetic energy by moving through a gravitational influence, we know that

$$\frac{E_{kinetic}}{m} - \Phi = 0, \tag{10}$$

where Φ is the gravitational potential, expressed as

$$\Phi = \frac{GM}{r}.$$

We can continue with substitution from equation (4) from chapter 56 into equation (10) provides this equation:

$$x\gamma v^2 = \frac{GM}{r},$$

where the values of x and γ are taken at r. With rearrangement,

$$v^2 = \frac{GM}{x_r \gamma_r r}.$$

This is a more accurate expression of velocity squared. It literally replaces the notion of a weak-field approximation with an exact value for velocity squared at distance r.

We can use this to explore gravitational redshift, which can be expressed as

$$z = \gamma - 1.$$

Substituting in the value of γ provides

$$z = \frac{1}{\sqrt{1 - \dfrac{v^2}{c^2}}} - 1.$$

And then substituting the exact value of velocity squared while normalizing the speed of light to a value of one provides a gravitational redshift equation of

$$z = \frac{1}{1 - \dfrac{GM}{x_r \gamma_r r}} - 1. \tag{11}$$

A photon traveling from infinity to a distance of r from a gravitating mass will experience the same change in energy-per-unit mass as an object traveling from infinity to r. This is the premise of gravitational redshift. This concept can be quantified using the Schwarzschild metric for a spherical, noncharged mass of uniform density.

$$ds^2 = \left(1 + \frac{\alpha}{r}\right)^{-1} dr^2 + r^2 \left(d\theta^2 + \sin^2\theta \, d\varphi^2\right) - c^2 \left(1 + \frac{\alpha}{r}\right) dt^2.$$

For an object falling radially inward, this simplifies to

$$ds^2 = \left(1 + \frac{\alpha}{r}\right)^{-1} dr^2.$$

And with further rearrangement,

$$\frac{ds}{dr} = \frac{1}{\sqrt{1 + \dfrac{\alpha}{r}}}.$$

The challenge now is finding the correct value for α. Traditionally, this is accomplished by applying the weak-field approximation through the following relationship:

$$\frac{d^2r}{d\tau^2} = \frac{\alpha}{2r^2}. \tag{12}$$

Equation (12) must reduce to Newtonian gravity for the limit of a weak gravitational field, so it is generally accepted that

$$\frac{d^2r}{d\tau^2} = -\frac{GM}{r^2}. \tag{13}$$

Therefore, through substitution and isolating α, one arrives at

$$\alpha = -2GM,$$

and

$$\frac{ds}{dr} = \frac{1}{\sqrt{1 - \dfrac{2GM}{r}}}, \tag{14}$$

when c is expressed in natural units with a value of unity.

Equation (14) is the basis of the GR gravitational redshift equation:

$$z = \frac{1}{\sqrt{1 - \frac{2GM}{r}}} - 1.$$

But let's go back and look at this substitution more carefully. The left side of equation (13) represents a change in velocity over time. The right side of equation (13) represents a force per unit mass. When you think deeply about this, you come to the conclusion that equation (13) can never be viewed as an absolute equality. That is, it is not possible to apply a constant force per unit mass (right side of the equation) and at the same time maintain a uniform change in velocity over time (left side of the equation). In fact, in every case except for the limiting condition in which there is no motion, the absolutely correct expression of equation (13) is

$$\frac{d^2r}{d\tau^2} < \frac{GM}{r^2}. \tag{15}$$

Equation (15) takes into account that an increased energy content decreases the change in velocity over time, as velocity increases under the influence of a constant force per unit mass.

This suggests that

$$\frac{\alpha}{2r^2} > -\frac{GM}{r^2}$$

IMPLICATIONS FOR GRAVITATIONAL REDSHIFT

and that

$$\alpha > -2GM.$$

One can attempt to reestablish an equality through relativistic force; however, once you recognize that the term $2GM/r$ represents a velocity squared, you immediately realize that the correct value of the term is $GM/x_r y_r r$ and that once modified, the GR equation is in complete agreement with equation (11).

58

RESOLVING THE DIFFERENCE BETWEEN GR AND SCALE METRICS REDSHIFT EQUATIONS

In the previous chapter, I suggested that gravitational redshift, z, can be determined through the following relationship:

$$\frac{ds}{dr} = \frac{1}{\sqrt{1 - \dfrac{GM}{x_r \gamma_r r}}} = 1 + z. \tag{16}$$

This is the GR equation for redshift adjusted for the actual values of x and γ for an object in free fall from infinity to a distance of r from a gravitating mass. This equation is solely based on a more honest treatment of kinetic energy and is completely independent of any influence of scale metrics.

There are those that will argue that $1/x_r \gamma_r$ reduces to two within the weak-field approximation and therefore that there is no contradiction with the GR redshift equation. Yet one must realize that the term $GM/x_r \gamma_r r$ represents a velocity squared, and one can express velocity squared in terms of γ as follows:

$$v^2 = \frac{\gamma^2 - 1}{\gamma^2}.$$

(17)

The derivation of this is as follows:

$$\frac{E_k}{m} = \gamma - 1,$$

and

$$x = \frac{\gamma}{\gamma + 1}.$$

Substituting the above into the relationship,

$$v^2 = \frac{E_k}{x\gamma m},$$

provides equation (17), which requires any positive velocity to be accompanied by a value of γ that is greater than one. Stated differently, if you insist on driving the value of γ to the limit of one, you must also drive velocity to the limit of zero. There can be no velocity without an increased energy content, and there can be no increased energy content unless γ is greater than one. This analysis is strictly by definition!

Let's next introduce scale metrics, where the redshift is determined by the ratio of the coordinate metric—which by definition is unity—over the proper metric, b, as follows:

$$\frac{1}{b} = 1 + \frac{GM}{r} = 1 + z.$$

(18)

RESOLVING THE DIFFERENCE BETWEEN GR

We now have two equations, (16) and (18), for gravitational redshift, and both should be correct. Does this represent a problem? Well, it actually presents an opportunity. If equation (16) (derived free of any influence of scale metrics) can be shown to be equivalent to equation (18), it provides significant support for the scale metrics model. The proof of this is provided below:

$$G = \frac{1}{2\pi h},$$

$$1 + \frac{M}{2\pi hr} = \left(1 - \frac{M}{2\pi hx\gamma r}\right)^{-\frac{1}{2}},$$

$$\left(\frac{2\pi hr}{2\pi hr + M}\right)^2 = \frac{2\pi hx\gamma r - M}{2\pi hx\gamma r},$$

$$\gamma = \frac{E}{M} = \frac{2\pi hr + M}{2\pi hr},$$

$$\left(\frac{2\pi hr}{2\pi hr + M}\right)^2 = \frac{(x)(2\pi hr + M) - M}{(x)(2\pi hr + M)},$$

$$\frac{(2\pi hr)^2}{2\pi hr + M} = 2\pi hr + M - \frac{M}{x},$$

$$x = \frac{E}{E + M} = \frac{2\pi hr + M}{4\pi hr + M},$$

$$\frac{(2\pi hr)^2}{2\pi hr + M} = 2\pi hr + M - \frac{M(4\pi hr + M)}{2\pi hr + M},$$

$$\left(2\pi hr\right)^2 = \left(2\pi hr + M\right)^2 - M\left(4\pi hr + M\right),$$

$$\left(2\pi hr\right)^2 = \left(2\pi hr\right)^2 + 4\pi hrM + M^2 - 4\pi hrM - M^2,$$

$$\left(2\pi hr\right)^2 = \left(2\pi hr\right)^2,$$

$$1 = 1.$$

This suggests that GR, when adjusted for the fact that $x_r \gamma_r$ is a variable across any interval, is in complete agreement with the predictions of scale metrics on the gravitational redshift of light.

This whole topic is made unnecessarily complicated by our use of a dual approach to kinetic energy. As stated in the opening paragraph of chapter 56, at low speeds we seem content with $E_k = 1/2\, mv^2$, and at higher speeds we simply apply a relativistic adjustment. The implication of this is that at low speeds, somehow the energy-per-unit mass lies in the velocity of an object. This has frequently been misstated even by physicists with comments such as "At low speeds energy goes into velocity, and at higher speeds it is manifested as mass."

This is a completely false statement. Energy is always manifested through an increased energy content (increased mass). Velocity is the result of an increased energy content. As the energy content increases, velocity will also increase. But this increase in velocity will occur at a slower and slower rate as more energy is added until the object ultimately approaches the limiting speed of light. This is because any additional energy that is added must not only move the invariant mass but also all the accumulated energy (as well as the orientation) associated with the increased energy content. The bottom line is that an object cannot possess velocity without an increased energy content. And this statement holds perfectly true across any interval, no matter how small. Yes, even at the interval of an infinitesimal!

Another way to look at this is that scale metrics introduces an additional variable that has not been taken into account within the traditional Schwarzschild metric. The idea of additional variables is nothing new and has been suggested by others, most notably Carl Brans and Robert Dicke. The Brans-Dicke theory is a metric theory of gravity that allows for a variable value of G.

The Brans-Dicke theory gained some popularity in the 1960s but eventually fell to the wayside in favor of GR. The fall of the Brans-Dicke theory was not due to its derivation but rather to the fact that to be in agreement with observation, the equation was generally reduced to that of GR, providing strong observational evidence that G is indeed a constant.

In scale metrics, it is not the case that G is treated as a variable but rather that the integer 2 from equation (14) is unlikely to be a constant across any interval (even within an infinitesimal interval within a weak field). Based on a careful and honest analysis of energy requirements, it is more accurately expressed as the variable $1/x_r y_r$, with the values of x and y changing across any interval. Once this is acknowledged, gravitational redshift is expressed as either equation (16) or (18), with both of these equations being completely equivalent.

59

WHY THE EVENT HORIZON NEVER FORMS

I n chapters 21 and 22, the idea was presented that the event horizon of a black hole never materializes at a set distance as defined by $r = 2GM$. To concisely make this point, based on equation (16) from the previous chapter, an event horizon should form under the condition where

$$x_r \gamma_r = \frac{GM}{r}. \tag{19}$$

Using the weak-field approximation with a velocity much less than c, the value of $(x_r \gamma_r)$ approaches the value of one-half, as required by the traditional GR redshift equation, but never reaches one-half. This subtle difference turns out to be extremely significant because there is no situation in which equation (19) is valid. That is, the variables x, γ, and r are all dependent on each other for any constant value of M. If one value changes, they all change. Because of this, there is no combination of allowable variables that satisfies equation (19) as an equality. The fact is that $(x_r \gamma_r)$ is never equal to $\left(GM \middle/ r \right)$ for any value of M and r. This inequality can be proven definitively.

Both y and x were exactly defined by equations (5) and (8). Based on GR, equation (10) establishes an equality between $\left(GM/r\right)$ and $\left(E_k/m\right)$.

Therefore equation (19) can be rewritten using terms E and m as follows:

$$\left(\frac{E}{E+m}\right)\left(\frac{E}{m}\right)=\frac{E-m}{m}.$$

This can be simplified to

$$E^2 = E^2 - m^2,$$

which is clearly an invalid expression for any case in which the space frame has mass, which is any case where $r > 0$.

When written correctly, equation (19) becomes

$$x_r \gamma_r > \frac{GM}{r}.$$

Or,

$$r > \frac{GM}{x_r \gamma_r}.$$

And this suggests

$$E^2 > E^2 - m^2.$$

If you wish to drive $\left(x_r \gamma_r\right)$ to the limit of one-half, you must also drive the value of r to the limit of zero. So an event horizon defined by $r = 2GM$ only occurs in the limiting case where the space frame—along with the value of r—goes to zero.

This can be shown graphically. Below is the case for the GR redshift equation suggesting $r = 2GM$.

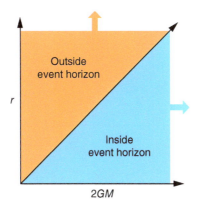

Since M and r are independent of each other, it follows that for any value of M there are an infinite number of allowable values for r. Those points in the orange region represent areas that lie outside the event horizon, and those points below the line in the blue region represent points within the event horizon.

But when we plot the same graph based on the relationship $r = \frac{GM}{x_r \gamma_r}$, we get much different results, as shown below. While M and r remain independent of each other, the value of $x_r \gamma_r$ is dependent upon both the values of M and r. Because of this relationship, all allowable values lie outside the event horizon.

The consequence of this is significant. Equation (19) precludes the formation of an event horizon for any values of M and r. Equation (16) never "blows up" with a value of zero in the denominator, and the concept of an event horizon becomes meaningless. The black hole as currently defined appears to be the result of an inaccurate application of classical mechanics incorporated into the derivation of the Schwarzschild metric. The resulting expression

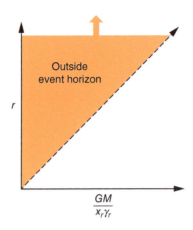

is an approximation that holds well in the range where $1/x_r\gamma_r$ remains reasonably close to the value of the integer 2. Beyond this range, the equation returns erroneous values that have little correlation with the physical world.

Does this claim hold up against observation? Well, it must be noted that tests of gravitational redshift are largely based upon Earth's gravitational field, which is a very poor producer of gravitational redshift. As stated in chapter 14, it is experimentally impossible to distinguish between GR and scale metrics redshift equations. Efforts to verify the agreement between GR and gravitational redshift observations have significantly improved over the past half century yet still fall far short of the necessary precision to differentiate between GR and scale metrics equations.

The Pound-Rebka experiment proposed in 1959 and conducted in 1960 verified the predictions of GR to within 10 percent of the observed gravitational redshift. In subsequent years, the precision was further refined to 1 percent. The next significant effort was Gravity Probe A in 1976 through the rocket redshift experiment performed by the National Aeronautics and Space Administration and the Smithsonian Astrophysical Observatory, which measured the actual change in the rate of atomic maser clocks separated by a distance of roughly ten thousand kilometers in Earth's gravitational field. GR was verified through this experiment to within one part per ten thousand. More-recent tests led by Holger Müller at the University of California, Berkeley, were based on an innovative experiment utilizing an atom interferometer that verified GR to seven parts per billion. The most recent effort was the Galileo gravitational redshift test with eccentric satellites in 2019. However, this experiment offered only modest improvements over Gravity Probe A. Future efforts have been proposed for additional space-based experiments, and it's possible the precision could be refined to within one part per hundred thousand billion (10^{-14}). However, none of these tests are precise enough to differentiate between the two redshift equations below:

$$z = \frac{1}{\sqrt{1 - \dfrac{2GM}{r}}} - 1,$$

$$z = \frac{1}{\sqrt{1 - \dfrac{GM}{x_r \gamma_r r}}} - 1.$$

The continued effort to improve observational limits serves as evidence that we are not assured of the absolute agreement between GR and gravitational redshift observations. In fact, for experiments utilizing Earth's gravitational field, the difference between the two equations above does not materialize until the eighteenth decimal place, or roughly one billion times beyond the most precise test to date. Therefore, observational data cannot be used to confirm or refute any differences between GR and scale metrics gravitational redshift equations.

However, what we do know is that any change in velocity requires a change in both x and y. Further, any change in x and y results in a change in g. It is impossible to hold g constant across any interval in which velocity is allowed to change. This holds for any interval, even at the limit of the infinitesimal interval. The only way to hold g constant is if the force per unit mass is increased across the interval. When applying this to gravitation, it would require the mass of the gravitating body to increase as the test particle falls through an interval, thereby ensuring a constant value of g across the interval. Further, this effect is cumulative, such that over a large distance the physical requirement would result in an ever-increasing mass of the gravitating body. And to achieve an event horizon at $r = 2GM$ would require an infinite amount of gravitating mass! As stated in chapter 21, nature had this covered all along. It takes an infinite amount of energy to reach the limiting speed of light, and it takes an infinite amount of mass to reach the limiting condition of an event horizon.

60

RELATIVITY: A CHEAT ON SCIENCE

E ven though this section is about mathematics, we are going to close the book with a more philosophical reflection. If all science is based on observation, then what happens if our observations are limited? I suggest we take a close look at the ramifications of this question on relativity theory (both special and general). While this chapter title is undoubtedly strongly worded, the point I wish to make is that relativity theory may well only hold true when we severely restrict the ability of an observer from making any observations of consequence. Are we then truly observing equivalence? Or are we simply imposing restrictions that make it difficult to observe differences (errors) due to the limitations of our measuring devices? Or are we possibly experiencing an inadequacy in our mathematical expressions to align with reality? If either of these prove true, then—yes—it is a bit of a cheat on science.

The final topic considered in this book is a significant one. It is the question of whether all motion is relative. Over the past century, most have considered this subject settled. Both relativity theory and general covariance clearly make this claim, but is it fully defended through observation? While many will suggest I am undertaking a steep uphill climb (perhaps even a foolish one), I nonetheless suggest we will find—in the end—that motion is subject to an absolute frame of reference, called space.

Let's start with special relativity (SR), the idea that two observers traveling with uniform velocities are unable to tell which is at rest and which is moving. This is indeed a true statement, and there is much to be learned from SR. It is an important piece in our overall understanding of nature. For me personally, SR was the entry point to my nearly half-century journey exploring gravity, gravitation, and the universe. However, there is a philosophical concern to be expressed. When you state that both observers undergoing uniform motion can claim to be at rest, it does not require that both observes have a valid point of view. The logic used is as follows: we have no way of telling who is correct (within the limits placed on our observations—i.e., uniform motion), and therefore they both have an equal claim on being correct (within their unique inertial frames of reference). This is not a self-evident conclusion, nor does logic dictate that this is the only conceivable outcome.

I grew up with two identical twins in my class at school. For the point of this discussion, let's say they were perfectly identical in every physical way. There was absolutely no way to tell the difference between the two based on appearance. Did this make them equivalent? Did I have an equally valid claim to use either of their names when I encountered one of them in the hallway with no fear of embarrassing myself or hurting their feelings? I can assure you that was not the case. They were two very different individuals with two unique sets of experiences. That I had no way of telling them apart based on physical appearance did not change the fact that they were two different people. To treat them the same would have been a cheat on reality.

Thinking more deeply about this, why was it that their family members could so easily tell them apart? Once I got to know them well enough (that is, observe them long enough), differences appeared through such things as mannerisms, sense of humor, and personality. Through observation I was able to tell the difference between two people who initially appeared to be the same. What was required? The ability to collect data over a long-enough period of time to document differences between the two.

Let's apply this to a universe that is roughly fourteen billion years old. I may see an object traveling with a uniform velocity, but it will not have been traveling at that same velocity over the entire life of the universe. That means if I can watch it long enough (into the future) or have a good knowledge of its path (from the past), I will eventually notice a change in either its speed, direction, or both. In other words, there will be a change in velocity over time in our relative observations.

And when this happens, using only the provisions of SR, I will have learned something about our two situations within space. For example, when a change occurs, I can ask myself whether I felt any force acting on me. If not, I can conclude that the object I am observing felt a force. I then know that the object I am observing accelerated and that I did not. I can repeat this analysis with every additional change that occurs. If I am allowed to fully document all motion for fourteen billion years, I will have established a frame from which all motion originally occurred. I can then document all motion as being relative to this initial frame (which is going to be my square of the universe).

This is a thought experiment that would be difficult to achieve because human life does not extend backward for fourteen billion years and because no one could maintain a line of sight with an object for that length of time. But that does not discredit its basic premise. Nature will indeed record each of these interactions as real physical events. Something real happens each time the velocity changes between two objects. This is something that Newton held firmly. He maintained a deep belief in an absolute frame of reference, even if it was only known to the mind of God. What humans may not be able to distinguish, nature clearly can.

The argument against what I have presented above is the theory of general relativity (GR). With the introduction of GR, relative motion was no longer restricted to uniform motion but had been extended to the more general (hence, *general* relativity) condition that accommodated acceleration. For the case of two observers, they could no longer assure themselves that they were at rest simply because they had not felt a force. It was now equally possible that either observer was in free fall moving toward the other observer who was at rest on the surface of a gravitating body. Therefore, the sensation of acceleration, which is entirely equivalent to gravity, could no longer indicate with certainty who was in motion and who was at rest. These ideas went to the heart of the Einstein equivalence principle and were central features in his development of GR. Today, scientists generally state this same idea by saying that in any small region of space and space-time the laws of physics hold true, including those of SR. This is the premise of general covariance, the idea that no coordinate system is better than any other (and—in fact—that they are all simply fabrications of the human mind (chapter 20).

Seems straightforward, but there may be potential problems lurking in the shadows. First, I am severely limited in my ability to observe my surroundings. Second, while the case is made that gravity and acceleration are equivalent, what is additionally required is an equivalence between gravitation and acceleration (chapter 19). Third, the established equivalence between gravity and acceleration is for uniform acceleration. What our two observers witnessed was a change in acceleration. Does this influence what they can observe? It might! Let's take these one at a time.

I am severely limited in my observations. I am only aware of a small moment in which a gravitating mass may be present and acting over a very tiny region of space. This is often modeled as an observer confined to a very small and sealed laboratory and possessing no knowledge of the outside world. In GR, we take all these small moments as experienced by observers in their labs (events), and we piece them together (through Riemannian geometry) to determine the smooth curvature of space and space-time. And this is perfectly legitimate as long as the mathematics represents something that may truly occur in nature. However, this is where a subtle distinction comes into play as to what is real. Specifically, what is an event? Within GR, an event is simply a set of coordinates (three spatial and one temporal). The question then becomes, How does the concept of an event relate to movement and the passage of time? When we use GR, time is treated as if it occurred through the sequencing of a large number of events.

Let's treat an event as a photograph of our twins. Since each photograph is only an instant, we are unable to recognize differences in personality, mannerisms, and the like from it. Our information is limited—even across time—to a sequence of instants. However, if we were allowed to fully view movement across time, we would have more information to tell them apart. This introduces an interesting situation. If we take all the photos—over very short intervals—and flip through them like a deck of cards, it will simulate motion and we will most likely be able to see the differences in mannerisms between our two twins. But this suggests that the act of observing is more complex than simply recording events. On the other hand, if the only thing that is real is the plotting of events, then any movement or passage of time that we perceive must be an illusion created by the human mind. It becomes the mind's way of ordering and processing a large number of instantaneous events.

When we look even deeper, we realize that within the framework of GR, time and distance are orthogonal to each other. By definition, this means they are completely independent of each other. That is, I can change the position along an axis of distance with no requirement to change a position on the axis representing time. The only way I can stay true to this requirement is if I limit my observations to the plotting of events. That is, my task is simply to record the coordinates that relate to each specific event. Neither the notion of movement nor the passage of time can ever be part of my physical reality because movement requires both a change in distance and time. Therefore, it simply does not exist. To allow it to exist breaks the orthogonal relationship between time and distance that serves as the framework of GR. If GR is correct, then what I experience as the flow of time is simply an illusion constructed within my mind.

Yet in the real world of physical science, where findings are based on observation, we all perceive the flow of time (as well as movement) to be real. In the case of our twins, the more time that goes by, the easier it is to tell them apart. We are not limited to an ongoing sequence of instants (events). So then ask the question: How small would an interval of time need to be for the mannerisms and personality differences between the twins to completely disappear?

This brings me to the second point. Even within the small confines of a tiny, sealed lab, there is a difference between acceleration and gravitation. Remember that gravity and gravitation are two different concepts in scale metrics. Gravity is a force, while gravitation simply defines the natural and preferred path of motion for an object moving within a gradient that defines a changing metric (chapter 19). The reality is that I can never make an interval so small that the difference between acceleration and gravitation disappears. Simply stated, I can never make a force equivalent to something that is not a force. I can mask it by reducing it to a level beyond the precision of my measuring devices, but I cannot make the reality of the difference disappear. I can incorporate the use of mathematics that enables the notion that acceleration and gravitation are equivalent. But this only ensures that my mathematics is consistent with my expectations. It does not ensure the mathematics is a true representation of what is physically possible.

Note that this line of reasoning is entirely separate from any consideration of tidal forces. Even within a theoretical rectilinear gravitational field where g remains constant with changing distance, I cannot create an interval small enough to eliminate the difference between gravitation and acceleration. I can say this with complete confidence because it is physically impossible for a change in velocity to occur over any interval at a constant rate of g without an increase in the force per unit mass—that is, an increase in the mass of the gravitating body.

For example, an object falling under the influence of gravitation will not travel the same distance as that of an object accelerating, in the absence of gravitation, at the same rate of g over the same time interval (chapter 19). And why is this? Because the object in free fall continues to gain energy content as it falls, and this additional energy makes future changes in velocity more difficult to achieve (assuming the mass of the gravitating body remains constant).

When we restrict our observation to an area so incredibly small (a tiny laboratory) that we cannot tell (measure) the difference, we must ask ourselves whether we are simply masking a difference or whether we are truly observing an equivalence between an object experiencing gravitation and an observer accelerating toward a stationary object in the absence of gravity?

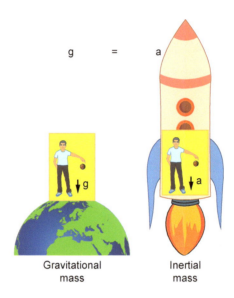

Consider this: am I unable to tell the difference between the twins because I have not observed them long enough or because they are the same? The answer to this is obvious.

For gravitation and acceleration to be equivalent requires the application of the EEP or general covariance, both of which suggest an object can gain velocity across an infinitesimal interval at a uniform rate defined by g. To state this is to assume that an object can gain velocity with no corresponding change in energy content. This is a central requirement of EEP, and general covariance—while stated differently—fully incorporates the EEP.

This can be conclusively shown to be false. Velocity can be expressed using a single variable, γ, as follows (chapter 58):

$$v = \frac{\sqrt{\gamma^2 - 1}}{\gamma}.$$

This expression conclusively shows that any velocity greater than zero requires a value of γ that is greater than one. If velocity is allowed to change over any interval, including an infinitesimal interval, the value of γ must also change across the interval. Further, any change in γ will result in a change in the rate at which velocity changes over time when subject to a constant force per unit mass. It is therefore physically impossible to hold a uniform value of g across an infinitesimal interval. If you wish to incorporate this concept into your mathematics, you must either apply a uniform rate less than g across the interval or use the true value of g while increasing the mass of the gravitating body as the object moves across the interval.

Since no one claims that the mass of a gravitating body increases with time, you can therefore conclude that one cannot hold g constant at a value equal to the force per unit mass across any interval, regardless of how small. If the force per unit mass remains constant, γ will increase and the value of g will decrease. There is no way around this!

In other words, it is a lot like trying to construct a perfect circle using line segments. This can be shown to work perfectly as long as your intervals are taken to the limit of an infinitesimal and as long as these infinitesimal intervals are truly defined by straight line segments. If you start with something close to—but not exactly—straight segments, you will end up with something close to, but not exactly, a circle (chapter 24).

Incorporation of the scale metrics equivalence principle highlights this point and shows this to be a cumulative effect that builds up over larger intervals as the values of x and y continue to diverge from the constant (classically treated) values of one-half and one:

$$v^2 = \frac{E_k}{x\gamma m} = \frac{GM_g}{x_r\,\gamma_r r} = \frac{M_g}{x_r\,\gamma_r\,2\pi hr}\,therefore\,\frac{ds}{dr} = \gamma = \frac{1}{\sqrt{1-v^2}} = \frac{1}{\sqrt{1-\dfrac{GM_g}{x_r\gamma_r r}}},$$

$$v^2 \neq \frac{2E_k}{m} = \frac{2GM_g}{r} = \frac{M_g}{\pi hr}\,therefore\,\frac{ds}{dr} = \gamma = \frac{1}{\sqrt{1-v^2}} \neq \frac{1}{\sqrt{1-\dfrac{2GM_g}{r}}}.$$

Furthermore, the inequalities shown above are true for all cases, including velocities that occur within the range of the weak-field approximation! I would suggest that there is no case in the real, physical world where the expressed inequalities can be established as an equality. Cannot happen!

Once this distinction is made, the case for absolute relative motion evaporates. Yes, motion can be relative for two observers in uniform motion. But that does not ensure that they both have a valid claim on the physical reality of their situations. Once a change in velocity is noted, something physical and real occurs.

Next, applying GR to this situation is not a complete solution. Motion can appear to be relative within uniform acceleration. But acceleration and gravitation have never been shown to be absolute equivalents; after all, one is a force and the other is not. Rather, all we can say with complete certainty is that they appear the same within the precision of our measuring devices over an infinitesimal interval or when we use mathematics that allows an object to gain kinetic energy with no increase in energy content. And what mathematics might that be? The use of a purely Newtonian expression within the weak-field approximation. Within classical mechanics, an object is not required to increase in energy content to possess a velocity greater than zero. When we use classical mechanics within the weak-field approximation, we embed a false truth into the mathematics of GR. We are suggesting a physical reality that we know is not possible to achieve.

Further, it is difficult to justify the use of an approximation when an absolute equation is available. The values of x and y do matter! We act as if we can discount this because— for example—a velocity of roughly 3 m/s corresponds to a value of x equal to 0.5000000000000000125 and a value of y equal to 1.00000000000000005. These values most likely lie beyond the precision of measurement, but that does not mean they are physically insignificant. Quite the contrary. The physical reality is that x is not one-half and y is not one when an object displays movement, and this reality lies at the core of the realization that kinetic energy cannot be achieved without an increased energy content, regardless of how small that change in energy content may be or the length or the time of the interval being considered.

In my work life, when I am not thinking about gravity and gravitation, I run a business. We have a saying in our office: pennies matter. We say this because it is true. It reminds me of a story—whether factual, legend, or just the stuff that movies are made of—about a programmer in the financial sector who rewrote code to shave a fraction of a penny off a great number of accounts. The idea of the scam was that the amount was way too small for any individual to notice but that collectively it could add up to millions of dollars. For an object moving at roughly three meters per second, are we not doing the same thing when we shave 1.25×10^{-16} off the value of x and 5×10^{-16} off the value of y? This may be too small for humans to recognize, but for nature, she will surely know. It may represent an arbitrarily small error across a tiny interval, but across time and through distance it will add up to a significant error. The result: the notion of an event horizon at a location of $r > 0$ (chapter 21).

This brings me to the third point, the difference between gravitation and acceleration when viewed from the perspective of a changing rate of acceleration. Remember, SR works for objects with uniform motion. GR claims to work within a uniform acceleration, which means I cannot tell whether what I feel is the result of acceleration or gravity. But what happens when we explore a changing acceleration? After all, this is the actual condition experienced by our two observers. They were not placed into an environment of uniform acceleration; rather, they were present to witness a change in acceleration.

So what happens when acceleration changes abruptly as observed from within the confines of a tiny, restrictive laboratory? If the change is due to acceleration within flat space-time (uniform distribution of mass and energy), the observer will be aware of an abrupt change in his or her weight. What happens if this change is due to gravity? The observer will also measure a change in weight but only after experiencing an impact. Prior to the moment in which a change in gravity is felt, the observer will have been following a natural and preferred path that resulted in an ever-increasing total energy content. Upon contact with the gravitating mass, the observer will experience an impact (where all the energy content due to kinetic energy is purged) followed by a change in weight. This experience is not identical to a change in acceleration!

Put another way, if I am present within flat space with no influence of gravitation, I can remain at rest in the same position for any extended period of time. After a significant amount of time has passed, the rocket in my laboratory fires, and I feel acceleration. I go from feeling weightless to experiencing weight over a short interval of time. Yet if I am in free fall in a gravitational influence, I will also experience a weightless environment during the entire interval of my free fall. However, in this case, before I experience gravity I first feel an impact. Once I recover from the impact, I will then feel my weight on the surface of the gravitating body. Think of it this way: When you fall out of a tree, it is not your weight that you are concerned about as you freely fall to the ground; it is the impact that you are about to experience. Impact will not occur if you simply apply a force to accelerate yourself at the rate of g.

If I feel no impact, I experienced a change in acceleration. If I do feel an impact, I experienced a change in gravity! Note that this still all occurred within observations made in my tiny laboratory over a small interval in time and with no knowledge of the outside world. I simply don't know when the impact will occur, but—when it does occur—it happens over a very short interval of time. Therefore, I have not violated the rules established for GR. Yet I have demonstrated a difference between a change in acceleration and a change in gravity. And why is that? Because I must be willing to take into account the energy content associated with motion. An object in free fall may not be aware of its motion, yet it is gaining energy as it follows its natural and preferred path. This energy must be purged (though impact) prior to claiming any equivalence between acceleration and gravity.

Within a pictorial description of scale metrics, we can visualize the case of an accelerating body, which is quite simple. An object at rest is placed within the uniform grid of the large-scale structure of the universe. If I wish to accelerate the object, I must force it to move in a manner against its natural and preferred path, which in this case is at rest. This applied force is felt as being equivalent to the sensation of gravity.

Now for the case of a gravitating mass: Within scale metrics, we would simply overlay a gravitating mass onto the large-scale structure of the universe. The gravitating mass emits free energimes outward from its center. Yet the gravitating mass itself feels no gravitational influence from the large-scale grid within your square of the universe. However, an object at a distance of r from the gravitating mass will experience a change in the metric as it moves radially inward toward the gravitating mass. Its movement is relative to the gravitating mass, which is positioned at rest within you square of the universe. Once it reaches the gravitating body, there will be an impact, and its movement will be terminated so that once again it is at rest relative to the large-scale structure of the universe. On the surface of the gravitating body, and after it has purged any energy content associated with its kinetic energy, the object will feel the force of gravity. The conclusion, once again, is that any change in gravity must first be accompanied by an impact prior to stating any equivalence between gravity and acceleration.

Now one could entertain a whole variety of thought experiments to discredit what I have said above. For example, what if my laboratory had been struck by an object, which then triggered its engines to fire? I would have felt an impact followed by a change in acceleration. Or what if I am on the surface of a gravitating body with a fixed mass that is collapsing inward at a constant rate? In this case, I would not feel any impact but would still experience a change in my weight. This is all very well, but I suggest that the individual in the laboratory in the real world is not likely to experience these one-off, extremely contrived scenarios. You may state that it does not matter—that they are possible, so therefore the theoretical equivalence can still be established. The point being missed is an emphasis on real-life observation. If my laboratory is allowed to document all changes

RELATIVITY: A CHEAT ON SCIENCE

over fourteen billion years, it may record an error here or there—due to the remote possibility of a counter-canceling phenomenon—but all that does is introduce some error bars as to what I will eventually determine as the absolute frame of space. It does not make motion relative; it simply introduces a level of error that would be no more significant than the probability of any of these unlikely events actually occurring.

To claim that no absolute frame exists is to claim that every time I feel an impact followed by a change in gravity, it will be due to a nongravitational collision of an object with my laboratory followed immediately by the correct firing of my laboratory engine to simulate the gravity that would be associated with the impact. I think we can easily eliminate this possibility from further discussion.

All three of the points I have made above are closely related. Time and motion both transpire over the course of any interval. While traversing that interval within a gravitational influence, the value of g changes as energy content increases. However, when moving across the same interval due to acceleration, the value of g is constant. This is required if you wish for gravity and acceleration to remain equivalent for both observers. And since gravitation and acceleration are not the same thing, any energy content gained because of kinetic energy (which drives the difference between gravitation and acceleration) must be purged before an equivalence can be established between gravity and gravitation. Of course this step can be ignored if you assume that it never happens. And you accomplish this by embedding classical mechanics into GR within the weak-field approximation. That is, if we use mathematics that ignores the increase in energy content over the interval, then we do not have to consider that it needs to be purged. The problem is that the weak-field approximation is wrong. We can use an exact expression for kinetic energy, velocity, or any other property by using the correct values of x and y within our calculations and equations. To assume that the classical values of one-half and one are ever correct is an invalid notion.

So how is this resolved using scale metrics? The answer lies in the space frame, the quantity of space (free energimes) that interacts with a gravitational influence at a distance of r. On a universal scale, the space frame is your square of the universe, and it represents all free energimes. These free energimes are real physical entities. They define the unity values of mass, distance, and time. This interval also defines the metric. Collectively, these free energimes establish an absolute frame of reference. Each energime possesses the property of motion, yet it is impossible to detect the movement of energimes within your square of the universe. That is because each energime is identical to every other energime.

Space is not a substance (ether) that matter moves through. Rather, space is the very fabric that couples with bound energimes to produce matter (chapter 36). This explains the null results of the Michelson-Morley experiment and raises an interesting perspective on the entire topic of wave-particle duality that is not addressed in this book. In addition, the energime segment introduces a new way to think about quantum action at a distance. All the "distance" defined by the energime segment indeed lies at the same location along the dimension of the segment. This is also a topic for further review not addressed in this book.

There is obviously more to consider, and the final chapter in the search for reality has not yet been written (and never will be). Having said that, perhaps the best way to end is with the concept of space...

Space, which defines the universe through an ongoing phase change from bound to free energimes; space, which defines the fundamental values of mass, distance, and time;

space, which exists as a physical entity defining the value of the metric as a real and tangible entity; space, which defines a space frame critical to the scale metrics equivalence principle; space, which encapsulates total energy content and precludes the formation of an event horizon; and yes, space, which perhaps defines the absolute coordinate system from which all events play out within.

If you end this journey with nothing else, walk away with the idea that space is far more robust than we have ever given it credit for. Space may hold many of the secrets yet to be unlocked.

It is space that will lead us to new insights.

ADDITIONAL READING

Brans, C. H., and R. H. Dicke. "Mach's Principle and a Relativistic Theory of Gravitation." *Physical Review* 124, no. 3 (1961): 925–35.

Brans, C. H. "The Roots of Scalar-Tensor Theory: An Approximate History." (June 10, 2005).

Delva, P., N. Puchades, E. Schönemann, F. Dilssner, C. Courde, S. Bertone, F. Gonzalez, A. Hees, C. Le Poncin-Lafitte, F. Meynadier, R. Prieto-Cerdeira, B. Sohet, J. Ventura-Traveset, and P. Wolf. "A Gravitational Redshift Test Using Eccentric Galileo Satellites." *Physical Review Letters* 121, no. 23 (2018): 231101.

Einstein, A. "Does the Inertia of a Body Depend Upon Its Energy Content?," *Annalen der Physik* 18 (1905): 639–41.

Hall, A. R., and W. J. 's Gravesande. *Dictionary of Scientific Biography*, Vol. 5, edited by C. C. Gillispie. New York: Charles Scribner's Sons, 1972. 509–11.

Hawking, S. W. *A Brief History of Time: From the Big Bang to Black Holes.* New York: Bantam Books, 1988.

Lightman, A., and R. Brawer. *Origins: The Lives and Worlds of Modern Cosmologists.* Cambridge: Harvard University Press, 1990.

Müller, H., A. Peters, and S. Chu. "A Precision Measurement of the Gravitational Redshift by the Interference of Matter Waves." *Nature* 463 (2010): 926–29.

Müller, H., A. Peters, S. Chu, N. Yu, M. Hohensee, B. Estey, F. Monsalve, G. Kim, P. Kuan, and S. Lan. "Gravitational

Redshift, Equivalence Principle, and Matter Waves." *Journal of Physics: Conference Series* 264 (2011).

Pagels, H. R. *The Cosmic Code: Quantum Physics as the Language of Nature.* New York: Bantam Books, 1983.

Pound, R. V., and G. A. Rebka Jr. "Apparent Weight of Photons." *Physical Review Letters* 4, no. 7 (1960): 337–41.

Pound, R. V., and G. A. Rebka Jr. "Gravitational Red-Shift in Nuclear Resonance." *Physical Review Letters* 3, no. 9 (1959): 439–41.

Pound, R. V., and J. L. Snider. "Effect of Gravity on Nuclear Resonance." *Physical Review Letters* 13, no. 18 (1964): 539–40.

Smoot, G., and K. Davidson. *Wrinkles in Time.* New York: Avon Books, 1993.

Schwarzschild, K. "Über das Gravitationsfeld eines Massenpunktes nach der Einsteinschen Theorie." *Sitzungsberichte der Königlich Preussischen Akademie der Wissenschaften* 1 (1916): 189–96.

Vessot, R. F. C., M. W. Levine, E. M. Mattison, E. L. Blomberg, T. E. Hoffman, G. U. Nystrom, B. F. Farrel, R. Decher, P. B. Eby, C. R. Baugher, J. W. Watts, D. L. Teuber, and F. D. Wills. "Test of Relativistic Gravitation with a Space-Borne Hydrogen Maser." *Physical Review Letters* 45, no. 26 (1980): 2081–84.

Unger, R. M., and L. Smolin. *The Singular Universe and The Reality of Time.* Cambridge: Cambridge University Press, 2015.

Will, C. M. *Was Einstein Right?: Putting General Relativity to the Test.* New York: Basic Books, 1986.

Zinsser, J. P. *Emilie du Chatelet: Daring Genius of the Enlightenment.* New York: Penguin Random House, 2007.

Zinsser, J. P. *La Dame d'Esprit: A Biography of the Marquise du Chatelet.* New York: Viking, 2006.

ABOUT THE AUTHOR

John R. Laubenstein has been an educator, researcher, and STEM education proponent throughout his career. He has taught physics, chemistry, and mathematics, conducted exploratory research at a major laboratory, and worked in the field of corporate philanthropy. He currently works with companies and not-for-profit organizations to manage charitable programs, many of which are focused on STEM education. His latest endeavor is sharing his lifelong journey exploring gravitation, relativity, and cosmology.

John completed his BS in chemistry at Northern Illinois University. During that time he started laying the foundation for what would become Scale Metrics. He believes challenging scientific topics can be discussed in meaningful ways without the need to delve into complex mathematics that often serve to drive away interested individuals.

John lives in Naperville, Illinois. He enjoys cycling and spending time with his children and grandchildren.

INDEX

absorption spectrum, 213-214
acceleration, 4, 7-8, 10, 89-91, 93, 95-98, 104, 238-239, 312-316, 318, 320-323
adiabatic expansion, 184, 186, 250
Aged, 171-173, 175-176, 188-189, 220, 278
air resistance, 3, 9, 88-89, 95
analogy, 20, 59, 82,195, 244
animation, 152
anthropic principle, 163, 165
antineutrino, 207
Apollo 15, 88-89
apple, 3, 5, 9, 11, 14, 16-17, 50, 56, 61, 65, 129-131, 133, 176
approximation, 67, 197, 282, 287, 292-293, 297, 303, 306, 318-319, 323
arc, 13, 125-126
at rest, 69-71, 73, 80,96, 100, 104-105, 109, 256, 290, 322
atom interferometer, 307
atomic maser clocks, 307
axis, 151-156, 314
background space, 26, 51, 56
balloon, 20, 185
BBs, 52

beta decay, 207-208, 210
big bang, 157, 164, 168, 191-193, 215, 249, 251, 255
black body, 250
black hole, 64-65, 74-75, 111-113,115-117, 139, 215-216, 303, 305
Bohr atom, 212, 220, 275
Bohr model, 211, 214, 275
Bohr radius, 277
bowling ball, 88
Brans, Carl, 301
Brans-Dicke theory, 137, 301
buffering, 27
calculus, 8
candy, 199-200
capture particle, 209
Cartesian coordinate system, 29, 56 151
cartoon, 152
celestial mechanics, 133, 137-138
CERN, 196
circle, 164, 177-178, 269, 317
circular orbit, 214
circumference, 45, 47, 53, 125, 178, 270
color bands, 213

compass, 125-126
Compton wavelength, 197-200, 205-206, 208-209, 220, 251, 264, 276
continuous spectrum, 213-214
coordinate distance, 14, 23,60
coordinate scale, 130, 187-189, 218, 225, 230, 236, 253, 271, 278, 281-282
coordinate time, 188-189, 230
cosmic microwave background radiation (CMBR), 159-161, 184, 191, 216
cosmological constant, 158, 162, 235
cosmological principle, 160-161
cosmology, 19, 159, 215-216, 223, 236-237, 250
Coulomb, Charles de, 49
Coulomb's constant, 43, 49, 276
coupling, 194, 198, 208-209, 215, 231, 251
curvature, 14, 17, 23, 25, 56, 92, 133-138, 313
dark energy, 162-163, 196, 218, 230-231, 233, 245-246
dark matter, 162-163, 196, 218, 230-231
degree of freedom, 155, 269
Dicke, Robert, 301
dimensionless, 44, 46, 139, 263-265
Doppler effect, 237-238
Doppler redshift, 237
doublet, 214
dough, 183, 238, 244
dry ice, 20, 185
du Châtelet, Émilie, 286
Eddington, Arthur, 137
eccentric satellites, 307

egg, 4
Einstein, Albert, 3, 10, 286
Einstein Equivalence Principle (EEP), 89, 92, 97-98, 102, 104-106, 114, 126, 137, 139, 316
Einstein radius, 112
elastic collision, 81-83
electric charge, 43, 49, 52, 54
electron, 46, 49-54, 63, 196-215, 220-221, 231, 233, 246, 250-252, 257, 261, 275-277
electron orbitals, 212, 276
electron spin, 211, 214
electrostatic, 49-52, 54, 63, 196-197, 203-206, 211, 215-216, 219, 246, 276, 281
elliptical orbits, 133, 214,
emission spectrum, 213-214, 275
energime, definition of, 35
energimes, bound, 44, 169, 178-179, 185, 188, 192, 194-196, 198, 200, 204-205, 215, 223, 231-233, 243, 246, 251, 281, 324
energimes, free, 37, 44, 46, 50-51, 55, 75, 107-109, 143-144, 146, 151, 154-155, 169-170, 177-181, 184-189, 192, 194-200, 204-206, 208, 211, 220, 223-226, 229, 231, 240, 243, 246, 251-252, 255, 268, 271, 281, 283, 322, 324
energy, kinetic, 65-68, 73-74, 81, 83, 91, 95, 98, 100-102, 104, 113-114, 118, 129, 138, 252, 262, 272-273, 285-291, 297, 300, 318-320, 322-324
energy, potential, 118
energy content, 66-69, 80-83, 85, 91-93, 95, 104, 126, 129, 138-139,

INDEX

264, 285-286, 288, 290, 294, 298, 300, 315-316, 318-323, 325
energy per energime, 221, 277-278
energy per unit mass, 220-221, 300
Eötvös, Loránd, 88
equilibrium, 44, 160, 230, 249, 251, 254
ether, 324
Euclid, 3
Euclidean geometry, 12-13, 15-16, 25-26, 79, 138, 255
Euclidean space, 8, 11, 29, 61, 178
event, definition of, 13
event horizon, 64, 111-112, 114-119, 139, 215-216, 256, 303-308, 320, 325
expansion, 67, 157, 161-163, 165, 168, 172, 175, 181, 184, 186, 189, 229-230, 233, 235-238, 250, 262, 281-282, 285
fabric of space, 143, 178, 237
feather, 89
fine-structure, 181, 193-194, 197-201, 205, 211, 214-216, 218-224, 232, 251, 257, 275-277
fishermen, 39-40
forest, 34-35, 155, 188, 232, 268
Friedmann-Lemaître-Robertson-Walker (FLRW), 282
fundamental mass, 46-47, 52-53
Galileo, 87
Galileo gravitational Redshift test with Eccentric sATellites (GREAT), 307
gamma, 68
general covariance, 92, 105-107, 109, 114, 126, 139, 262, 309, 312, 316
general relativity (GR), 10, 13, 16-17, 21-24, 33, 46, 55-63, 74,

89, 92, 96, 106, 111-114, 117, 119, 126, 133-139, 149, 152, 156-157, 178, 183, 223, 237-240, 245, 262, 279, 294-295, 297, 300-308, 312-314, 318, 320-323
Geneva, 196
geodesic, 13-14, 92
global, 12, 76, 84, 121-123, 272
God, 24, 312
Grand Canyon, 117
gravel truck, 82
Gravesande, Willem 's, 286
gravitation, 16, 18, 49-50, 54, 61-63, 75, 87, 93, 95-98, 101-102, 104, 106, 111, 115, 126, 133, 137-139, 143-144, 181, 203-204, 211, 217, 231, 233, 245-246, 256-257, 261, 269-270, 281, 290, 308, 310, 312, 314-316, 318-320, 323
gravitational mass, 87-88, 90, 114
gravitational potential, 291
gravity, 3, 7-10, 15-17, 19, 24-25, 30, 49, 51-52, 59, 63-64, 89-91, 93, 95-97, 102, 104, 106, 133, 137-139, 143, 157-158, 162, 203, 246, 256, 261, 290, 293, 301, 306-307, 310, 312, 314, 319-323
Gravity Probe A, 306-307
Guth, Alan, 161
hammer, 89
Heisenberg Uncertainty Principle, 109
Higgs boson, 196
Hilbert, David, 112
homogeneous, 160-161, 164, 181, 218, 230, 256
Hoyle, Fred, 191
Hubble, Edwin, 158

Hubble, 160-161, 165, 167-168, 237-239, 257
Humason, Milton, 158
identical twins, 310
illusion, 151-152, 314
inelastic collision, 81-82
inertial mass, 87, 89, 90, 95, 97, 104, 316
infinitesimal, 8, 92, 95, 97-98, 102-103, 114, 119, 121-123, 126, 137, 139
infinity, 12, 16-17, 56, 63-64, 114, 129-131, 133, 204-205, 292, 297
inflation, 161, 163-166, 218, 227, 229
initial conditions, 161
initial singularity, 191-192, 158-159, 215, 250
instantaneous, 7, 98, 100-101, 134, 314
integrated wave-particle duality, 109
interval, 8, 19, 46, 60, 92, 95, 98, 102, 108, 119, 126, 129, 134, 137, 139, 145-146, 153, 173, 216, 223, 256, 300-301, 308, 314-324
inverse fine-structure constant (number),
isothermal, 185-186, 254
isotropic, 151, 160-161, 164, 181, 218, 229-230, 256
kernel of corn, 98-99, 243
laboratory, 184, 313, 315, 320-323
language, 4, 152, 255, 262
large-scale structure of the universe, 23-27, 51, 56, 61, 75, 109, 134, 136, 138, 231, 268, 321-322
Laws of Nature, 105-106, 217

leaves, 232, 241
Legoland, 121
Lorentz factor, 287
Lorentz transformation, 68
lunar atmosphere, 89
Mach, Ernst, 23
Mach's principle, 23, 61
mass, inertial, 87, 89-90, 95, 97, 104, 316
mass, relativistic, 66, 74
matter, formation of, 210
medley, 150
mesh, 144, 169-170, 185, 187, 195
Michelson-Morley experiment, 324
Milky Way, 24-25, 27, 115, 210
model-based observation, 237, 239-241, 249, 251
momentum, 8, 76, 79-85, 101, 107, 131, 214, 270-274
motion vector, 77, 84-85, 108-109, 272
Müller, Holger, 307
multiverse, 163-164
napkin, 173-175, 198-199
NASA, 306
natural and preferred, 10, 89-90, 96, 100, 102, 104, 135-136, 139, 314, 320-321
Nature, 39-41
neutron, 54, 196, 203, 207-211, 231, 257
Newton, Isaac, 3
Newtonian, 65-66, 118, 293, 318
Nobel Prize, 162
number theory, 44
parade, 199
parallel universe, 163
particle motion, 109
Penzias, Arno, 159

INDEX

periodic table, 207, 211
Perlmutter, Saul, 162
Peter Higgs, 196
philosophers, 65, 151
photon, 44, 75, 77, 135-136, 197,
 221, 240, 263-265, 272, 277-278,
 292
picnic, 198-199
Ping-Pong ball, 88
Planck, Max, 38
Planck distance, 46-47
Planck mass, 46-47
Planck scale, 38
Planck time, 46
Planck's constant, 38, 43-45, 53,
 226, 263-264, 267, 270
polar coordinates, 105
popcorn, 243
positron, 205-206, 208-209, 233
Pound-Rebka, 306
Professor Now, 171-176, 189
property of motion, 71, 98, 138,
 324
proton, 46, 53-54, 196, 201, 203,
 205-212, 215, 220, 246, 257, 276
Pythagorean Theorem, 17, 56, 101,
 130-131
quantum mechanics, 208, 212
quantum model of gravitation, 54
quark, 210
raisin bread, 183-184, 238, 244
rate of phase change, 154, 188
reality, search for, 5, 258, 324
rectilinear gravitational field, 315
relativistic physics, 66
rest mass, 66-70, 72, 80, 91
Riemann, Bernhard, 3, 12
Riemannian geometry, 12, 102,
 114, 121, 127, 313

Riess, Adam, 162
rocket, 23, 306, 320
rocks, 154
role of science, 217
roller coaster, 9
rubber band, 183-184
rubber sheet pictorial of gravity, 59
Rubik's Cube, 150
ruler, 244
Rutherford, Ernest, 203
Sagittarius A*, 115
SAO Rocket Red Shift Experiment,
 306
Scale Metric Equivalence Principle,
 87, 93, 99, 101-102, 268, 273,
 317, 325
scale metrics, 16-17, 19-25, 44,
 46, 50, 53, 56-57, 62-64, 71, 74,
 76, 84-85, 87, 89, 92-93, 95-96,
 98-102, 106-121, 126, 129, 134,
 137-139, 147, 149-156, 168-170,
 177-178, 184-185, 187, 192, 194,
 196-197, 200, 203-204, 208, 211,
 215-216, 218-225, 229-231, 233,
 235-236, 240-246, 249-278, 281,
 283, 297-301, 306, 308, 314, 317,
 321-325
scaler, 84-85, 131
scaling factor, 85, 101-102, 107,
 261, 270-271, 273-274
science, 3-5, 9, 15, 30, 51, 64,
 99-100, 111, 117, 162, 216-217,
 237, 253, 257-258, 262, 309, 314
scientists, 5, 7, 11, 13, 15, 38, 65,
 74, 83, 92, 144-146, 151-152,
 162-163, 165-166, 184, 193, 216,
 221, 237-238, 245, 312
Schmidt, Brian, 162
Schwarzschild, Karl, 112

Schwarzschild radius, 112
Scott, Dave, 89
segment, 29, 33-34, 100, 125-126, 138, 143, 155, 173, 177, 188, 232, 261, 268-269, 317, 324
self-regenerating particle (SRP), 205-206, 208-209, 233
shed, 116
shotgun shell, 52
singularity, 20, 158-159, 165, 169, 191-192, 215, 243, 250, 281
six-foot boxes, 72-73
Smithsonian Astrophysical Observatory, 306
snow, 224-226
solar system, 24-25, 27, 115-116, 133
Sommerfeld, Arnold, 211
space frame, 101, 104, 261, 267-268, 270, 304, 324-325
spaceship, 89, 91, 96-97, 102
special relativity (SR), 96, 99, 310-312, 320
spectrum, 185, 212-214, 219, 237, 250, 257, 275, 281
sphere, 13, 76, 92, 146, 177
standard model, early, 158-165, 219, 229
standard model, 156-165, 168, 172, 174, 181, 183-184, 191, 196, 203, 207, 210, 216, 219-221, 223, 229-231, 235-240, 243-244, 249-250, 252, 278, 281
static property of matter, 71, 138, 256
steady bang, 192
steady state, 191-192
string theory, 34
strong nuclear force, 54

sublimation, 20, 185
supernova, 218, 239
Switzerland, 196
Taylor series, 67, 73, 285, 287
test particle, 60-61, 63, 96, 98, 100-104, 129, 308
thermal signature, 192, 252
thought experiment, 82, 311, 322
three-foot boxes, 72-73
tidal forces, 315
time, arrow of, 149, 151
time, flow of, 76, 151, 153-154, 314
time, meaning of, 151
time, passage of, 70, 144, 188-189, 193-194, 245, 313-314
time, special moment in, 165-166, 168, 239, 257
time-like, 56
torsion balance, 88
trap door, 118
Traveler, 173-176, 179, 187
tree branch, 3, 129-131, 133
trees, 34-35, 155, 188, 232, 268
Type Ia supernova, 218
ultraviolet, 213
universal constant, 193, 216, 265, 267-268
universe, age of the, 30-31, 37, 46, 164-165, 167-168, 172-173, 175-176, 189, 198, 218, 220-223, 225-227, 229, 238, 257, 263, 275, 278
universe, birth of the, 173, 178-179
wave motion, 109
wave-particle duality, 109, 324
weak field approximation, 292, 297, 303, 318-319, 323
weak nuclear force, 210
Wheeler, John, 14

INDEX

Wilson, Robert, 159
window screen, 144, 195
windowpane, 107, 144, 185, 195
Woodshop, 102-103, 127
yo-yo, 205

Young, Thomas, 65
your square of the universe, 154,
178, 181, 187-188, 218, 229,
246, 264-265, 271, 281-282,
324

Lightning Source UK Ltd.
Milton Keynes UK
UKHW052235300821
389744UK00006B/97/J